Brain-Grounded Theory of Temporal and Spatial Design

Yoichi Ando

Brain-Grounded Theory of Temporal and Spatial Design

In Architecture and the Environment

Springer

Yoichi Ando
Kobe University
Kobe, Japan

ISBN 978-4-431-55889-7 ISBN 978-4-431-55891-0 (eBook)
DOI 10.1007/978-4-431-55891-0

Library of Congress Control Number: 2015961026

Springer Tokyo Heidelberg New York Dordrecht London

Printed on acid-free paper

Springer Japan KK is part of Springer Science+Business Media (www.springer.com)

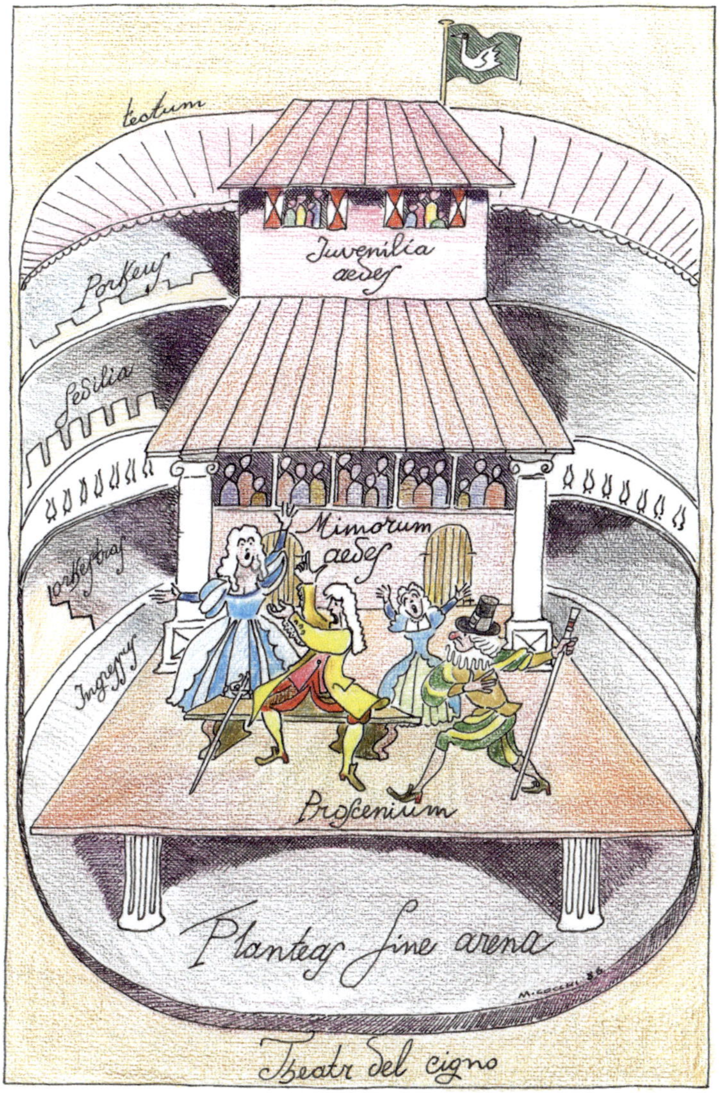

When different individual personalities blossom, the wonderful results they produce continue to contribute to the quality of human life in unfathomable ways for a long time even after the end of the individual life. Each individual is born with such special potential and talent for a unique third stage of life.

The drawing (1986) was rendered by Massimo Cocchi, a man who loved music and enjoyed words at a certain level of loudness, and who realized clarity of the sound field without the use of ceiling reflections or multiple sidewall reflections.

Foreword

In my opinion, even if he was already involved in many research fields, all converging towards the best knowledge of the nervous system as a means of communication between man and nature, when in 1995 Prof. Ando invited many researchers in the field from all over the world to visit his new realization, the theatre in Kirishima, he already had in mind the future development of his scientific interest in a more complete approach of architecture to the environment. The opportunity for realizing this meeting was the organization of a new international conference, called at that time Music and Concert Hall Acoustics.

As a matter of fact, to that conference were invited not only acousticians but also an architect from the United States who spoke of light and time as elements to be taken into consideration during the architectural drawing course. Another element that promoted the idea of a wider interest was the location of that theatre and its external shape, conceived like a leaf fallen in a glade among the trees.

Finally, I realized that this was Yoichi Ando's target only in 2001 when he decided to publish a new scientific journal devoted to temporal design (*Journal of Temporal Design;* JTD). Quite soon after (2002), I received the honour of a personal meeting with Prof. Ando in Rimini (Italy) during which he revealed to me his idea with so many details that it was quite impossible not to take a deep interest in his scientific project.

In 2003 he called the world of scientific research to Kobe to achieve the first meeting of the International Symposium on Temporal Design (1st ISTD). Now it is common to hear of the ISTD every 2 years in some place in the world, as a confirmation of the quality of the scientific work done by Yoichi Ando in the field of interactions between man and the environment.

Statements that it is possible to find in this book are all supported by scientific work done in cooperation with physiology and neuroscience experts or personal studies carried out in different opportunities of his life: for instance, the writer of this short foreword was involved as a subject of measurements of the human reaction in the face of acoustical and visual stimuli.

Readers looking for a light lecture about some "Oriental" discipline of easy learning will not be at their ease from the beginning of this book. If, however, they are guided by the mental approach of those who deeply desire to understand something new about links between human life and the environment in which we are involved, surely they will find in it many responses to their scientific curiosity.

As an engineer, I always search for new ideas that will improve human buildings in the direction of realizing an indoor artificial living environment, in enjoyable conditions, comfortable in the way that a natural environment is, with full respect to a natural environment and natural resources. In this context I have particularly appreciated the sentence "A general strategy for design is to characterize what humans experience and what they prefer to experience (preferences), and to optimize their environments so as to realize their preferences" in Yoichi Ando's presentation of temporal design at the Science Academy of Bologna (July 2015).

There are many natural situations that induce nervous stimuli easily reproducible only on a perceptive basis while designing artificial environments, but there are also others that require deep knowledge of physical, not arbitrary, rules. In my opinion, temporal design theory can also develop in this direction in the near future, searching for the best equilibrium between stimuli and their physical origin.

For instance, a typical example can be found in Paragraph 9.3, where we read: "The most effective spatial factor to consider with temperature is vertical distribution along the body, particularly below the knee". This is a very important sentence, calling for deep research into the best distribution of surface temperature in the walls surrounding a person and its links with energy consumption—a fundamental physical problem to be seriously considered and resolved for the future of our planet. In that respect, this book is not only a milestone but also an opening for future research.

Bologna, Italy
August 2015

Alessandro Cocchi

Preface

From a brain-grounded theory of design based on subjective preference of the sound field in concert halls (Ando 1985, 1998, 2007, 2009a, b) *to a brain-grounded theory of temporal and spatial design in architecture and the environment, our research affirms that by encouraging personal traits and satisfying individual needs, subjective preference works as an approach to enhancing individual health and creativity. Most generally, subjective preference is regarded as the primitive response of a living creature that entails judgments that steer an organism in the direction of maintaining life, so as to enhance its prospects for survival. Therefore, subjective preference may be deeply associated with the base of aesthetics.*

The auditory preference model consists of two kinds of internal representations of sound that are based on the autocorrelation- and crosscorrelation-like structures of sound as it presents itself to the two ears. The autocorrelation function (ACF) describes the monaural and temporal signal at each of the two ears, while the interaural crosscorrelation function (IACF) describes the binaural and spatial signals arriving at the entrances of the two ears. Remarkably, temporal sensations, such as loudness, pitch, duration, and timbre, are well described by the temporal factors extracted from the ACF of the sound signal; and spatial sensations, such as localization, apparent source width (ASW), and subjective diffuseness, are described by the spatial factors extracted from the IACF (Ando 2009a, 2015).

The information corresponding to subjective preference of sound fields was found in the effective duration of the ACF of the alpha waves, registered by both the electroencephalogram (EEG) as well as the magnetoencephalogram (MEG). The repetitive feature of the alpha wave, as measured in its ACF, was observed at the preferred condition. This evidence suggests that the basic theory of subjective preference may be applied to individual preference as well. By analyzing the MEG, we reconfirmed that the left cerebral hemisphere is associated with the temporal factors of the sound field, and the effective duration of the ACF alpha wave directly corresponds to the scale value of individual subjective preference. The right cerebral hemisphere is activated by the typical spatial factors, i.e., the magnitude of interaural crosscorrelation (IACC) and the listening level.

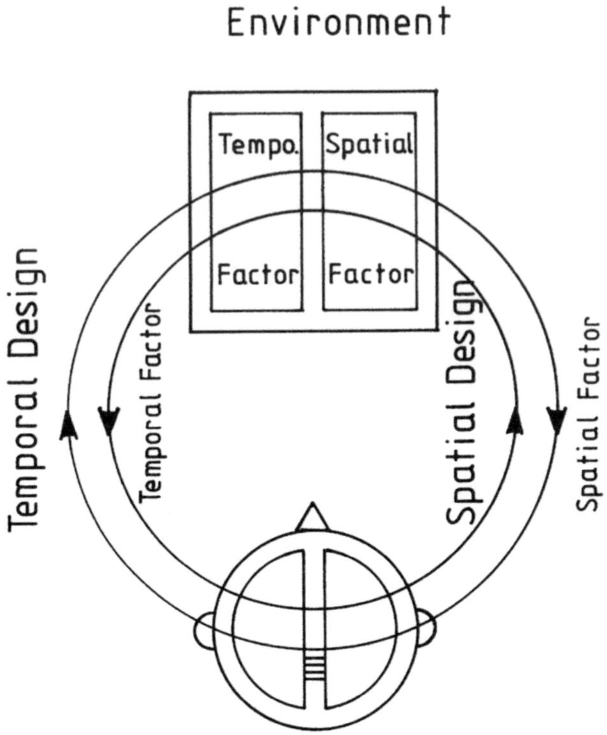

Fig. P.1 Interaction between environment consisting of temporal and spatial factors and the left and right human cerebral hemispheres, respectively, associated with the temporal and spatial factors. This is the ground of theory for designing the architecture and the environment as well as artistic works. Such a well designed environment realized by the subjective preference theory induces further creations developing individual personality (Ando 2009a)

The theory of subjective preference has been developed with auditory perception in mind; it can plausibly be extended to predict subjective preferences in analogous dimensions of visual perception. Thus, parallels can be drawn to temporal and spatial sensations in vision as well as to subjective preferences for visual environments. For example, the most preferred condition of a flickering light is expressed by the temporal factors extracted from the temporal autocorrelation function (ACF) of the modulated light stimulus. It is shown how the temporal and spatial factors of auditory and visual percepts are associated with the left and right human cerebral hemisphere, respectively.

A brain-grounded theory of temporal and spatial design has been proposed as indicated in Fig. P.1 (Ando 2009b).

Three Stages of Human Life (Time): A Definition

For temporal design of the human environment, three stages of life which are granted to each individual person at birth are considered in this volume. The first of these is the physical life for which the main nutritive element is food, the second is mental life for which nutritive elements are culture and art including, for

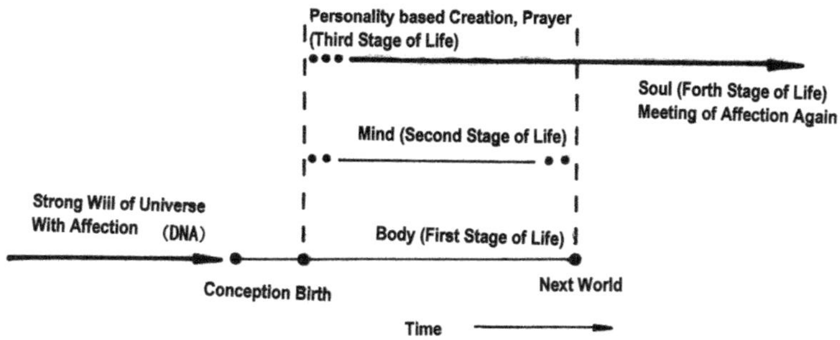

Fig. P.2 Three stages of human life defined in this volume that are taken into consideration in planning the temporal and spatial environment (Ando 2006).
Human life consists of: (1) life of body, (2) life of mind, and (3) life of creation, which is based on the unique personality of the individual that persists long after the individual has passed on. Thus, the third stage of life includes creations based on personality together with awaking higher perception, which continues to the fourth stage of life after passing the life (Fig. P.3). The third life is a source of joy that broadens a life from a philosophical standpoint, making it unique to humans compared to the life of animals who only enjoy first and second lives. Nourishing the first and second stages only inevitably induces "ill treatment" and further "wars" similar to animal life

example, music, painting, literature, and fashion. These two stages are what humans share with animals; the only difference lies in respective proportions. The third stage is distinctive of humans as shown in Fig. P.2 see also Fig. P.3; "the third life" is also considered the noblest of the three. According to individual DNA and the varying environmental conditions in which we grow up, each of us may have a unique purpose to find and work on throughout our lifetime (Fig. P.4). The elements that either nourish or prevent the growth and development of this stage are the limitations of art and science at the present stage, worship and philosophy, and degree of conscious design in the environment where individuals develop to find - or miss - their "mission". It is possible to achieve health-promoting conditions on all three levels of an individual's life by the temporal design of the environment.

It can be argued that science, technology, and art are forever immature (Fig. P.5) since they are always under development. The nineteenth and twentieth centuries are commonly viewed as the era in which technology made rapid progress. However, such progress came at the high price of environmental destruction and war. In hindsight we can say that it was overconfidence in our knowledge of science and technology that brought about such disastrous results; science and technology should be regarded as mere means for mankind, not as ends in themselves. In order to make a coming age of fulfilling human life a reality, we must understand mankind itself better; this can be achieved by incorporating the concept of the third stage of life into our thinking process. On this account, we should prepare an environment where every individual's unique sensitivity is regarded and valued as the groundwork for a meeting place between science and art.

Fig. P.3 "Full of affection of universe" painted by Keiko Ando (2015)

The question of how many years mankind will be able to survive is currently the primary international and social concern. There have been major wars under the plea of racial liberation, religious liberation, or the liberation of a state, and the

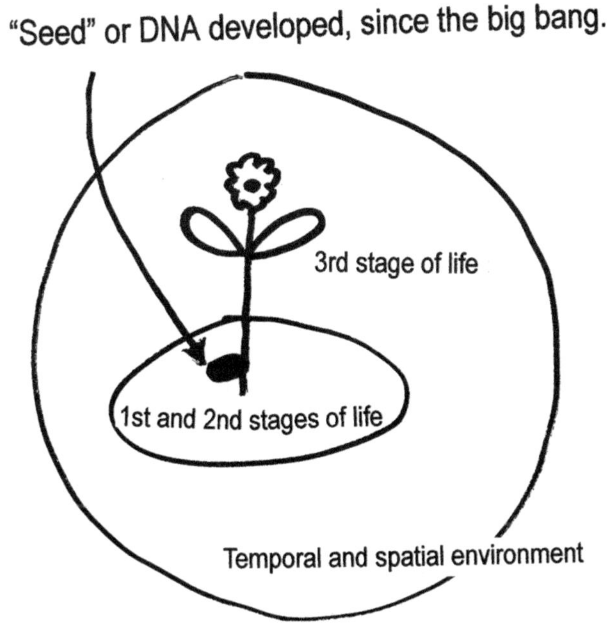

Fig. P.4 Development of the third stage of human life originating from a genetic "seed" that is nurtured in an appropriate temporal and spatial environment. In a preferred environment, the personality for a unique creation develops and effloresces like the flower of a healthy plant

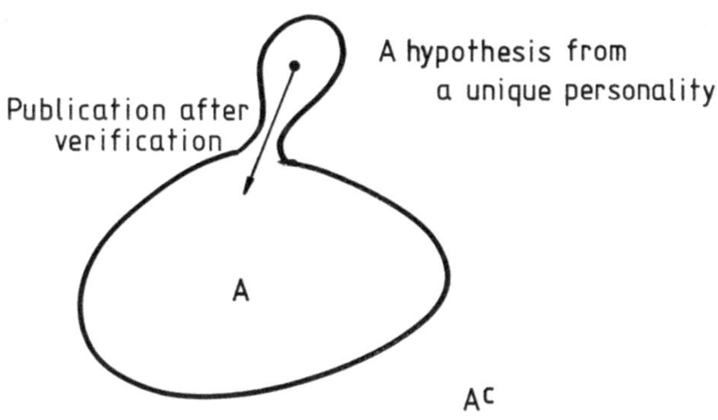

Fig. P.5 Sets of the known A and the unknown A^C. Creations start with a hypothesis from a unique personality. After verification of the hypothesis, the result may be published. This is a process of the third stage of life. Integration of these kinds of creations is called "culture"

likelihood of such events reoccurring cannot be denied at present. Therefore, we hope to make the liberation of unique personalities (the third life) the ultimate objective of this volume. The liberation of individuality here means to accept and value the diversity of individual traits in every person, and to cultivate creativity, which can be achieved only by supporting each individual potential.

A well-designed environment could be a meeting place for art and science and may help to discover individual personality as the minimum unit of society. Temporal and spatial factors associated with the left and right cerebral hemispheres, respectively, may be well-used in design that blends a built environment together with Nature. In turn, such environments may support the development of the human cerebral hemispheres, especially during the period from the very beginning of life to about 3 years of age.

Role of Individuals in Society

Space (land) was once thought to be a personal possession, and time was thought to be the same for everyone in the past. But in fact space (including the environment) is something common to all mankind from generation to generation, and fundamentally, time is that which belongs to the individual because of the nature of life. It is ideal to build an environment as well as a society where people can enjoy healthy physical, spiritual and individual time for creation.

Above all, we would like to establish that the possibility to nurture creativity (to make a contribution generation after generation) as it presents itself in individual time is unique to mankind, as is the possible liberation of individuality. In other words, there are mysteries solvable only by each person's individuality due to differences in the DNA and the emergent viewpoints, which will never be generated again (Fig. P.6). Such unique solutions may gain their eternal value only when unveiled, and it is no exaggeration to say that those mysteries unsolved are much like solutions that remain unfound as a result of uniform education.

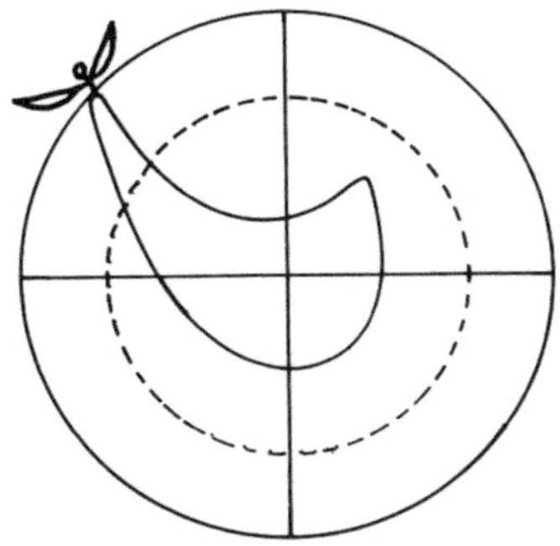

Fig. P.6 Individual potential steered by the personality (DNA "seed") in a particular direction to discover something original that has not been previously known. For example, typical directions are music, arts, athletics, mathematics, science, engineering, literature, medicine, history, geophysics, agriculture, economics, business, and other fields to be developed by individual personalities. The *dotted circle* in the figure depicts an individual conforming too much to the accepted norm, thus not discovering anything new and original. For example, competitions are not a good environment for the development of the third stage of life (e.g., "honor students") because they place too many restrictions on what is acceptable

Mysteries of this sort exist at different points of every human activity. To provide a place for all individuals to nurture a task that has been given to them until a solution sprouts out of their heart-mind should be the starting point of education. We believe that this kind of education will be ultimately established, so that the human and global environment (time and space) may contribute to a healthy development of culture and science. For these reasons, it is ideal to liberate the original individuality from all idol worships of mere economic efficiency, status, reputation, and so forth. Such individual liberation will not only mark a process towards peace as it entails mutual respect between individuals and diversity groups, but also holds hidden potential for becoming the path to protect all lives from environmental destruction. In order to develop individual personality and creation, we are attempting to provide a means for the temporal design of our environment for the third stage of life in addition to the first and second lives, as a hope for the sustenance and perseverance of life. This may be a base for the flourishing of each individual and development of a healthy society and environment.

It is worth noting that the dimension of the head of newborn babies is relatively large because this part is first developed in the body. If we consider an extended analogy, it is highly recommended that the facilities most closely related to the human brain be first designed in house and urban planning, such as museums, concert halls, libraries, churches, and any institutions that may be important for the development of the third life. The arms and legs, corresponding to highways and communication systems, may be developed later on. Because of the continuing global climate change, a dynamic method for the temporal design of the environment that blends human life with the natural environment should be considered.

Previous Major Publications Related to This Topic

Ando Y (1985) Concert hall acoustics. Springer, Heidelberg

Ando Y (1998) Architectural acoustics, blending sound sources, sound fields, and listeners. AIP Press/Springer, New York

Ando Y (2007) Concert hall acoustics based on subjective preference theory. In: Rossing T (ed) Springer handbook of acoustics. Springer, New York

Ando Y (2009a) Auditory and visual sensations, Peter Cariani (ed). Springer, New York

Ando Y (2009b) Theory of temporal and spatial environmental design. In: McGraw Hill 2009 yearbook of science & technology, pp 384–389

Ando Y (2015) Opera house acoustics based on subjective preference theory. Springer, Tokyo

Soeta Y, Ando Y (2015) Neurally based measurement and evaluation of environmental noise. Springer, Tokyo

Kobe, Japan Yoichi Ando

Acknowledgments

The epitome of this volume was presented at a Commemorative Lecture Meeting held at the Academy of Sciences in Bologna on July 16, 2015, in memory of appointment of the author as a member of the academy. The meeting was organized by Professor Alessandro Cocchi, University of Bologna, in the field of Architecture. After the lecture meeting, Marianne Jogi and the author discussed her PhD project in the study of Alessandro for about one hour and afterwards, we all visited "Santuario di San Luca" on top of a hill near Bologna as shown in Photos P.7 and P.8.

In the comprehensive yet accurate English expressions found in this volume, Marianne Jogi, an artist and a Ph.D. candidate from Tallinn University of Technology, Estonia, has conveyed individual and cosmic rhythms while also exemplifying in Fig. 9.6 how it is possible to incorporate temporal aspects in environmental and architectural design to support "the third life".

Fig. P.7 After the lecture meeting on July 16, 2015, the author (*left*), Marianne Jogi (*center*) and Alessandro Cocchi (*right*) visited the church of "Santuario di San Luca", a landmark of Bologna, situated on top of a hill near the city

Fig. P.8 "Santuario di San Luca", a drawing (1986) rendered by Massimo Cocchi, a man who enjoyed Bologna during his life

This volume serves as a record of the research performed at the Ando Laboratory, Graduate School of Science and Technology, Kobe University. For warm cooperation in research work since 1970 at Drittes Physikalisches Institut, University of Goettingen, the author thanks Professor Manfred Schroeder, Dr. Peter Damaske, and Dr. Dieter Gottlob. In 1975 the author received an Alexander von Humboldt Fellowship for his research work on concert hall acoustics.

Shin-ichi Sato, Hiroyuki Sakai, Yoshiharu Soeta and Kenji Fujii who received their Ph.D. degrees at the graduate school of Kobe University, performed a number of excellent research works, and their publications are cited here.

The author would like to express his appreciation to the Committee members of Hillside Residence Planning including Thomas Bosworth as a member of the small group in the Architectural Institute of Japan. As a result of prolific discussions, we published a book entitled *Hillside Residence Planning Blending with Natural Environment, Incorporating Temporal and Spatial Factor*s (Gihoudou, Tokyo, Japan; 1995).

Drs. Akira Fujimori, Shioko Okada, and Shoko Ikeda, Konan Hospital, Kobe, oversaw continuous medical treatments to maintain the health of the author.

Abbreviations

ASW	Apparent source width, one of the spatial sensations, which is described by the spatial factors extracted from the IACF of the sound field, Sect. 4.1.2
ACF	Autocorrelation function. Temporal sensations may be described by the temporal factors extracted from the ACF of the sound signal
NACF	Normalized ACF, Sect. 4.2.1
DS	Duration sensation, which is introduced here as one of four temporal sensations, Sect. 4.2.1
EEG	Electroencephalogram, Sect. 2.2
FF	Fundamental frequency [Hz], F_0, which is the measure of pitch, (see also τ_1) Sect. 4.2.1
IACC	Interaural crosscorrelation coefficient, magnitude of the IACF, the most significant consensus factor among the four orthogonal factors of the sound field, Sect. 4.2.1
IACF	Interaural crosscorrelation function. The spatial sensations of the sound field may be described by the spatial factors extracted from IACF by analyzing the sound signals arriving at the two ear entrances, Sect. 4.1.2
LL	Binaural listening level, [dBA], or binaural sound pressure level measured by the geometric mean of $\Phi_{ll}(0)$ and $\Phi_{rr}(0)$, Sect. 4.2.2
MEG	Magnetoencephalogram, Sect. 2.2
PCT	Paired-comparison test (Thurstone 1927; Gullikson 1956; Torgerson 1958): Most of the subjective preference judgments and other subjective responses in this volume were conducted by the PCT. Usually, trials started with a first stimulus, followed by a short blank duration and then a second stimulus. During the subsequent blank duration, the subject judged which stimulus was the subjectively preferred stimulus. The scale value is related to the probability of whether stimulus A is preferred to B. For example, if $P(A > B) = 0.84$, then the value is 1.0. The value, therefore, may be reconfirmed by the goodness of fit (Mosteller 1951). All of the data in this volume was reconfirmed by the test. This shows that the model

of obtaining the scale value was approved. The scale values of the subjective judgments of each individual subject can also be calculated (Ando and Singh 1996; Ando 1998). If the experimental procedure is identical, then the probability data may be integrated over time and space, because such a comparative method is simple and easy to use on anyone, even children, to obtain good reliability and thus good reproducibility. Thus, recording of the SVR, EEG and MEG were performed in a manner similar to the paired-comparison method to find the relationship between the factor extracted from the correlation analysis of the recorded signal and the scale value of subjective preference judgments.

PLG Plethysmogram; photo-electrically observed at the fingertip as a pulse wave. The peripheral blood vessel reacts as a reflection of the autonomic nervous system (the sympathetic system) and this may be observed in the PLG. Section 14.3

SV Scale value obtained from the PCT. The value given by S may be described by the temporal and spatial factors

SVR Slow vertex response as an auditory evoked potential, reflected activities on the left and right hemispheres (latency is less than 500 ms), Sect. 2.1

Contents

Comments

[1] Prof. Dr. Nag Woon-Hae : Japanese, Korean missionary dispatched by the United Church of Christ in Japan, Co-Pastor of Saemoonan Presbyterian Church in Seoul, Assistant professor at the Presbyterian University and Theological Seminary in Korea, Visiting professor at the Seigakuin University General Research Institute.

We should guide our attention to the individuality of the "third stage of life" that is advocated in this volume. Professor Ando's scientific individual-based approach offers a completely new insight.

Professor Ando's original and creative ideas engage in a holistic manner with the deeply wounded, groaning and ailing world – a vision that encompasses not just mankind, but everything that is "related to life", such as animals and plants, the earth, air, sound, and even history which is the axis of time, thereby allowing us to sense the creative evolvement of a new form of knowledge.

Moreover, Professor Ando's language choice includes terms that have for a long time been issues of theological debate: "life", "time", "love", "freedom", "individual", "creation", "the last day", etc. Many of the terms he has selected have great theological significance. I hope to be able to learn from Professor Ando's scientific mind.

We theologians in Korea are preparing for a new theological movement, namely the 'Ohn Theology – Holistic Theology'. It is a movement emphasizing the importance of theology as a holistic discipline. Interdisciplinary discussions carry great significance in this movement. The reason behind this is the belief that theological development can occur only when we venture beyond the boundaries of Christian theology to engage in discussions with other religions and other academic fields, including science.

[2] Chief Priest of Saikoji Temple, Kohji Danjo.

There are many kinds of conflicts and clashes in the world now. Many of them come from strife among religions, races, nationality, economy and so on. And many people have been injured, killed, suffered loss, and roamed in strange countries.

Religions originally make people happy. But the differences between religions cause serious problems that many people suffer from. However, it is impossible to integrate many religions.

We must pay attention to a greater world, if we can regard “Something Great” as the “Will of Cosmos”, a sphere of humanity beyond all differences.

On the other hand, generally speaking, many people are apt to attach importance to money, economy, and the material, the so-called visible world. But, it is desirable that a greater number of people be interested in the invisible world as “life” or “mind”, I think.

In the scientific field, the connection between material and mind has been discussed and newly reported in quantum mechanics and near-death experience.

Quantum mechanics considers both “the coincidence of mind and body” and ”the coincidence of consciousness and material” to be important. Our consciousness is thought to be connected with a sub-consciousness and universal Will.

Should we consider how our individual consciousness connects with cosmic consciousness? Many scientists in the world have been studying and reporting on near-death experiences, and there is a common element to be recognized. These studies seem to suggest a present world and another world. Gratitude is important for all people in the world now. Could our world be improved by deepening our connection between the first (body) and second (spirit) stage and the third and fourth stage of human life?

Chapter 1
Introduction

1.1 Dimension of the Environment: Outer Universe

It is well known that the physical universe (environment) consists of one-dimensional time and three-dimensional space. Mathematically, Hermann Minkowski (1864–1909) defined the physical distances as

$$s^2 = {x_1}^2 + {x_2}^2 + {x_3}^2 + {x_4}^2 \tag{1.1}$$

where $x_1 = x$, $x_2 = y$, $x_3 = z$, and $x_4 = \text{ict}$, i being $(-1)^{1/2}$. This signifies the four-dimensional continuity.

Let us focus our attention on enabling each individual granted access to the third stage of life by considering a human brain grounded theory of temporal and spatial design in architecture and the environment as shown in Fig. P.1.

1.2 Dimension of Subjective Preference: Inner Universe

The typical overall response of the inner universe is the cooperation between the two cerebral hemispheres. The result of this unified level of functioning manifests itself as the most primitive response of the subjective attributes of an individual, which is known as subjective preference. Preferences guide the organism in the direction of maintaining life; thus we may observe brain activities related to preference. Preference is also deeply related to an individual's aesthetic sense. The point of contact between art and science may, therefore, help to discover individual preference or personality as the minimum unit of society or the public. If individuals are satisfied by the environment, then the public is satisfied, but the converse is not true. This volume emphasizes individual life. Maximum individual

Y. Ando, *Brain-Grounded Theory of Temporal and Spatial Design*,
DOI 10.1007/978-4-431-55891-0_1

satisfaction for the approximate number of 7.26×10^8 people in our world should be rather easy to accomplish with the help of computers.

The paired-comparison method has been used to obtain the scale values of subjective preferences and the actual related brain activities have been found. Acoustic design of a concert hall and an opera house maximizing the preference scale values at each seat has been explicitly described. After construction of a concert hall, the seat selection system, which was first introduced at the Kirishima International Music Hall in 1994, may help to enhance individual satisfaction. On that note, an opera house and a theater may be argued to represent a miniature version of the environment of our everyday lives, because “All the world’s a stage” (William Shakespeare, 1564–1616).

The theory of subjective preference for the sound field has been described by use of the temporal and spatial factors extracted from the respective correlation function mechanisms existing in the neural system (Ando 1985, 1998, 2007, 2009a). The left and right cerebral hemisphere specialization associated with temporal and spatial factors, respectively, may play an active role in determining the relationship between independent effects and preferences. The scale value of subjective preference of the visual field has been described by both temporal and spatial factors (orthogonal). Consequently, a brain-centered theory for planning the environment, which incorporates both temporal and spatial factors, is proposed (Ando 2009b). For example, the scale value of subjective preference S as a distance of the sound field is expressed by

$$S = S_1 + S_2 + S_3 + S_4 \tag{1.2}$$

where S_1, S_2, S_3, and S_4 each represent the scale value of subjective preference for each of the four orthogonal factors of the sound field (Eq. 1.2). This is the four-dimensional continuity similar to Eq. (1.1). The theory of subjective preference of the sound field has been extended to encompass visual fields (Ando 2009a, 2015), whereas other modalities (thermal, vibration, and smell) are yet to be investigated with regard to the specialization of the human cerebral hemispheres.

1.3 Three Stages of Human Life

For the design of architecture and the environment, we shall discuss a neurally grounded theory that is based on subjective preference. The design of the sound field in a concert hall (Ando 1985, 1998) and opera house (Ando 2015) will be applied for this purpose. We are proposing an approach called “temporal design” in addition to spatial design in the field of architecture and the human environment (http://www.jtdweb.org/2001; Ando et al. 1996; Ando 2004, 2009b, 2013).

With regard to this approach, what can be seen as the three stages of human life have been identified (Fig. P.2). We propose that all useful creations that continue to contribute to human life even long after the first and second lives have passed are based on the unique and healthy personality of the individual. Therefore, we recommend that the human environment be designed with respect to all three stages of life, as environment is plausibly a most important factor in the development of an individual's personality. In turn, a well-designed environment may contribute to the development of the brain. Temporal and spatial factors associated with the left and right cerebral hemispheres, respectively, as shown in Fig. P.1, can be incorporated into design that blends built environment and nature. Such environments close to newborn babies may support the development of the human cerebral hemispheres, especially during the period before attending kindergarten and/or elementary school, by which time almost the entire brain will have formed.

1.4 Twilight of the Third Stage of Life of the Author: An Example

One day when the author was around 10 years old, the teachers of the elementary school decided to take the class to the movies. The theater was located in a town about 5 km from the school. The entire class was extremely excited to be sitting in a theater that could house an audience of about 200 people. The ceiling was wavy in the longitudinal section, and someone said that it must be the result of calculations to obtain effective scattered reflections. However, he began to wonder how exactly it was calculated and evaluated the elegance of the complicated surface structure. He forgot about the title and the story of the movie entirely and pondered the puzzle for years to come.

About 24 years later, in 1963 after joining the Architectural Acoustics Laboratory at Kobe University, he began investigating an interference pattern method of measuring the complex reflection coefficient for an oblique sound incident for a flat wall surface in an anechoic chamber. A circular horn pipe about 20 mm in diameter attached in front of a loudspeaker radiated the pure tone between 2 and 15 kHz. A probe microphone consisting of a circular pipe 2 mm in diameter attached to a static (condenser) microphone was moved to obtain the interference pattern (Ando 1968a). In order to obtain accurate values of complex reflection coefficients of materials, the acoustic centers (location of point source or receiving point) and directivities of both the circular horn loudspeaker and the probe microphone with a circular pipe were measured and reconfirmed by calculations (Ando 1968b, 1969/1970).

An opportunity arose to attend the international congress on acoustics held in Budapest in 1971, and the author visited the Third Institute of Physics, University of Goettingen, where he worked for about 1 month in the summer. There, with Peter Damaske, the author investigated a method of calculating the interaural crosscorrelation function, which is related to subjective diffuseness (Damaske and Ando 1972).

At Kobe University, the author recalls experiencing a wonderful sound field, which speech signals came from reading a poem when the direction of the single reflection was adjusted to about 30° in the horizontal direction and the delay time was 20 ms (Ando and Kageyama 1977).

The doctorate degree was not easily obtained, because at that time the author was performing research work with Professor Hiroshi Hattori, School of Medicine, Kobe University, on the long-term effects of noise on unborn babies around the Osaka International Airport (Sect. 14.3: Ando and Hattori 1977a, b). The dispute was based on evidence at the Osaka Local Court of Justice where the Japanese government and the people living close to the airport were arguing with a group of about 20 lawyers. The author (Yoichi Ando), a doctoral candidate at that time, received a telephone call from an acoustician working at the Research Laboratory of Science and Technology of NHK (Japan Broadcasting Corp.), who was backed by a professor at the Institute of Space and Astronautical Science, University of Tokyo (founded by the Japanese Prime Minister in 1877), and officers from the Ministry of Transport, Japan. During the phone call, he was asked to cancel his upcoming talk at the Court of Justice where he was scheduled to introduce the evidence he had gathered in his own work. The author, however, gave evidences on effects of aircraft noise on unborn babies. As a result of the incidence, the author could not obtain the degree from a national university, where he had applied by submitting his doctoral thesis. After about 5 years, fortunately, he received the degree in 1975 by invitation from Professor Takeshi Itoho, Waseda University (private university founded in 1882 by Shigenobu Okuma).

Soon after, he was invited by Manfred Schroeder, the director of the Third Institute of Physics in Goettingen, as an Alexander von Humboldt Fellow to investigate room acoustics. Just before the invitation, he had performed the groundwork with K. Kageyama at Kobe University; subjective preference judgments were conducted using the simplest sound field with a single reflection by changing its directions and delay time. The consideration was that with such a simple sound field, which is representative of infinite sets of reflections, it might be possible to solve the long-time puzzle of what is, in fact, the most preferred sound field for humans (Ando and Kageyama 1977). Here, subjective preference was selected as an evaluation of the sound field, because it is the most primitive response in subjective attributes and the evaluative judgments. Preferences guide the organism in the direction of maintaining life. In humans, therefore, these preferences are deeply related to aesthetic issues. Typical results are shown in Fig. 1.1a as a function of the delay time of the single reflection. At Goettingen, the author discovered that the most preferred delay times correspond well to the effective duration (τ_e) of the autocorrelation functions of two different music motifs. For music motif A (slow tempo: Royal Pavane composed by Orland Gibbons, $\tau_e \sim 127$ ms), the most preferred delay time was 128 ms. And for music motif B (fast tempo, Sinfonietta, Opus 48; movement IV, $\tau_e \sim 35$ ms), the most preferred

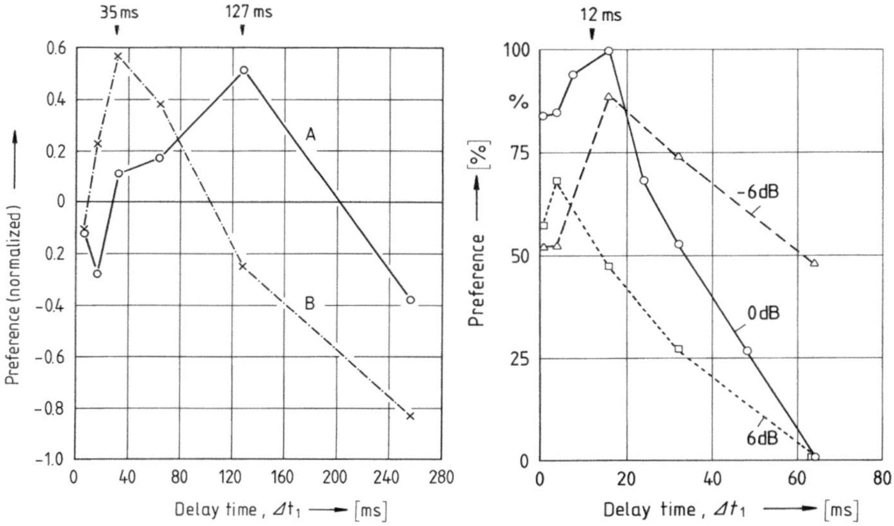

Fig. 1.1 Preference score as a function of the delay time of the single reflection
(**a**) Music Motif A (Royal Pavane by Gibbons, $\tau_e \sim 127$ ms) and Music Motif B (Sinfonietta, Opus 48, III movement, by Arnold, $\tau_e \sim 35$ ms)
(**b**) Speech, a female reading a poem, $\tau_e \sim 12$ ms, scores are plotted as a parameter of the amplitude of reflection
It is remarkable that the most preferred delay times of the reflection correspond well to the value of τ_e (the effective duration of the ACF of source signals), when the amplitude of reflection is the same as that of the direct sound.
The concept of "temporal design" sprung from the results, because the sound field of rooms may be designed according to the purpose of the room, i.e., for speech or for slow/fast music

delay was 32 ms (Fig. 1.1b). Note that for speech signals with $\tau_e \sim 10$ ms, the longest delay time was about 10 ms. For quite different types of music, the preferred horizontal directions were similar, 32°–64°, and the corresponding amplitudes of the interaural crosscorrelation, the IACC, were at minimum in the range of 50°–60° as shown in Fig. 1.2 (Ando 1977). Thereafter, preference tests were continued on the sound field with multiple reflections and reverberation time (Ando and Gottlob 1979; Ando 1983).

Later, the author found that subjective preference as an overall response is a cooperation between four orthogonal factors – two temporal and two spatial factors associated with the left and right hemispheres, respectively (Ando 1985, 1998) – and went on to establish the theory of subjective preference with temporal factors associated with the left hemisphere and spatial factors associated with the right hemisphere including the visual fields (Chap. 5, Ando 2007, 2009b).

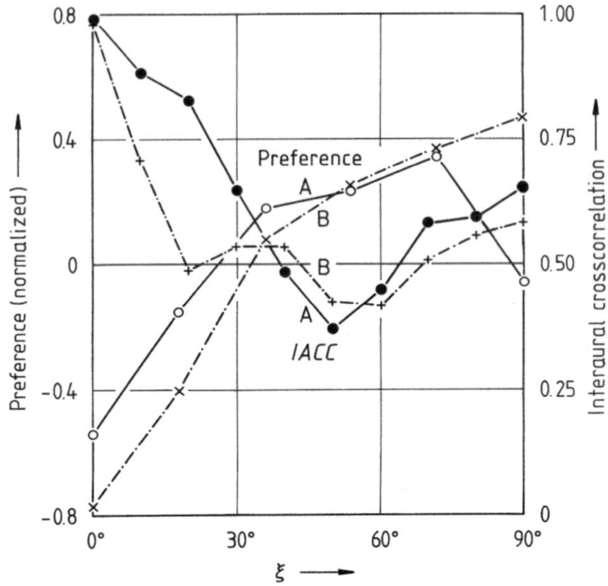

Fig. 1.2 Preference scores and magnitude of the interaural crosscorrelation (IACC) of the sound fields, as a function of the horizontal angle of the single reflection ($A_1 = 1$) The IACC indicates minimum when horizontal angles of the reflection are around 55° at which preference values are relatively large. This is the basic concept of spatial design of a room, which is almost independent of sound signals

References

Ando Y (1968a) The "interference-pattern method" of measuring the complex reflection coefficient of acoustic materials at oblique incidence. Electron Commun Jpn 51-A:10–18

Ando Y (1968b) Experimental study of the pressure directivity and the acoustic center of the circular pipe horn loudspeaker. Acustica 20:366–369

Ando, Y. (1969/1970). On the sound radiation from semi-infinite circular pipe of certain wall thickness. Acustica 22:219–225

Ando Y (1977) Subjective preference in relation to objective parameters of music sound fields with a single echo. J Acoust Soc Am 62:1436–1441

Ando Y (1983) Calculation of subjective preference at each seat in a concert hall. J Acoust Soc Am 74:873–887

Ando Y (1985) Concert hall acoustics. Springer, Heidelberg

Ando Y (1998) Architectural acoustics, blending sound sources, sound fields, and listeners. AIP Press/Springer, New York

Ando Y (2004) On the temporal design of environments. J Temporal Des Architect Environ 4:2–14. http://www.jtdweb.org/journal/

Ando Y (2006) Reviews on the temporal design for three stages of human life. Most Unlikely "time is money, "but" time is life". J Temporal Des Architect Environ 6:2–17. http://www.jtdweb.org/journal/

Ando Y (2007, 2014). Concert hall acoustics based on subjective preference theory. In: Rossing TD (ed) Springer handbook of acoustics. Springer, New York

Ando Y (2009a) Auditory and visual sensations. Springer, New York

Ando Y (2009b) Theory of temporal and spatial environmental design. In: McGraw-Hill yearbook of science & technology 2009. McGraw-Hill, New York, pp 384–389

Ando Y (2013) Environmental design for the third stage of human life (persistence of individual creations). J Temporal Des Architect Environ 12:1–12. http://www.jtdweb.org/

Ando Y (2015) Opera house acoustics based on subjective preference theory. Springer, Tokyo

Ando D, Gottlob D (1979) Effects of early multiple reflection on subjective preference judgments on music sound fields. J Acoust Soc Am 65:524–527
Ando Y, Hattori H (1977a) Effects of noise on sleep of babies. J Acoust Soc Am 62:199–204
Ando Y, Hattori H (1977b) Effects of noise on human placental lactogen (HPL) levels in maternal plasma. Br J Obstet Gynaecol 84:115–118
Ando Y, Kageyama K (1977) Subjective preference of sound with a single early reflection. Acustica 37:111–117
Ando Y, Johnson B, Bosworth T (1996) Theory of planning environments incorporating spatial and temporal values. Memoirs of Graduate School of Science and Technology, Kobe University, 14-A, pp 67–92.
Damaske P, Ando Y (1972) Interaural cross-correlation for multichannel loudspeaker reproduction. Acustica 27:232–238

Chapter 2
Model of Auditory Brain System

2.1 Slow-Vertex Responses (SVR) Corresponding to Subjective Preference of the Sound Field

Neural response correlates of subjective preference were found in the latency of SVR peaks. Figure 2.1 summarizes the relationship between subjective preference scale values and three acoustic parameters (SL, Δt_1, and IACC). Applying the paired method of stimuli, both the SVRs and the subjective preferences for sound fields were investigated as functions of these parameters (Ando 1998). The source signal was 0.4–0.9 s of segment. The lower part of the figure indicates the appearance of latency components.

1. As shown in the left and center columns in the figure, the neural information related to subjective preference appeared typically in an N_2-latency of 250–300 ms, when SL and Δt_1 were changed. In general, subjective preference may be regarded as a predisposition towards maintaining life, making the appearance of such primitive survival-oriented responses in neuronal observables all the more expected.
2. Further details of the latencies for both the test sound field and the reference sound field, when Δt_1 was changed, are shown in Fig. 2.2. The parallel latencies at P_2, N_2 and P_3, were clearly observed as functions of the delay time $\Delta \mathrm{t}_1$. However, latencies for the reference sound field ($\Delta t_1 = 0$) in the paired stimuli were found to be relatively shorter, while the latencies for the test sound field with $\Delta t_1 = 25$ ms, the most preferred delay, became longest. This pattern may indicate a kind of *relative* behavior of the brain, which leads to underestimating the reference sound field when the test sound field in the pair is the most preferred condition.
3. Relatively long-latency responses are always observed in the subjectively preferred range of each factor.

Y. Ando, *Brain-Grounded Theory of Temporal and Spatial Design*,
DOI 10.1007/978-4-431-55891-0_2

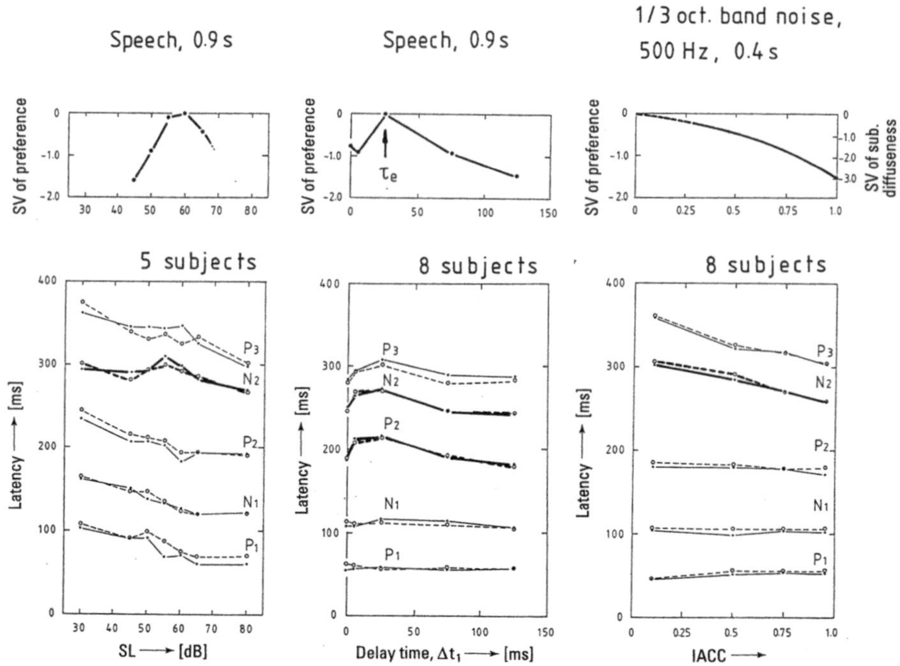

Fig. 2.1 Relationships between averaged latencies of the SVR and scale values of subjective preference for three factors of the sound field
(____): left hemisphere; (------): right hemisphere
(**a**) As a function of the sensation level (*SL*)
(**b**) As a function of the delay time of reflection, Δt_1
(**c**) As a function of the IACC

4. Thus, the difference of N_2-latencies over both hemispheres in response to a pair of sound fields contains almost the same information as that obtained from PCTs for subjective preference.

   The right column of Fig. 2.1 shows the effects of varying the IACC using the 1/3-octave-band noise (500 Hz) (Ando 1987b, 1992). At the upper part, the scale value of the subjective diffuseness is indicated as a function of the IACC. The scale value of subjective preference also has a similar behavior plotted against the IACC.
5. The information related to subjective diffuseness or subjective preference, therefore, appears in the N_2-latency, ranging between 260 and 310 ms, in which a tendency for an increasing latency while decreasing the IACC was observed for eight subjects (except for the left hemisphere of one subject). The relationship between the IACC and the N_2-latency was found to be linear, and the correlation coefficient between them was -0.99 ($p < 0.01$).
6. Furthermore, let us look at the behavior of early latencies at P_1 and N_1, which remained almost constant when the delay time and the IACC were changed. However, the information related to the SL may be found typically at the N_1-latency. This tendency agrees well with the results of Botte et al. (1975).

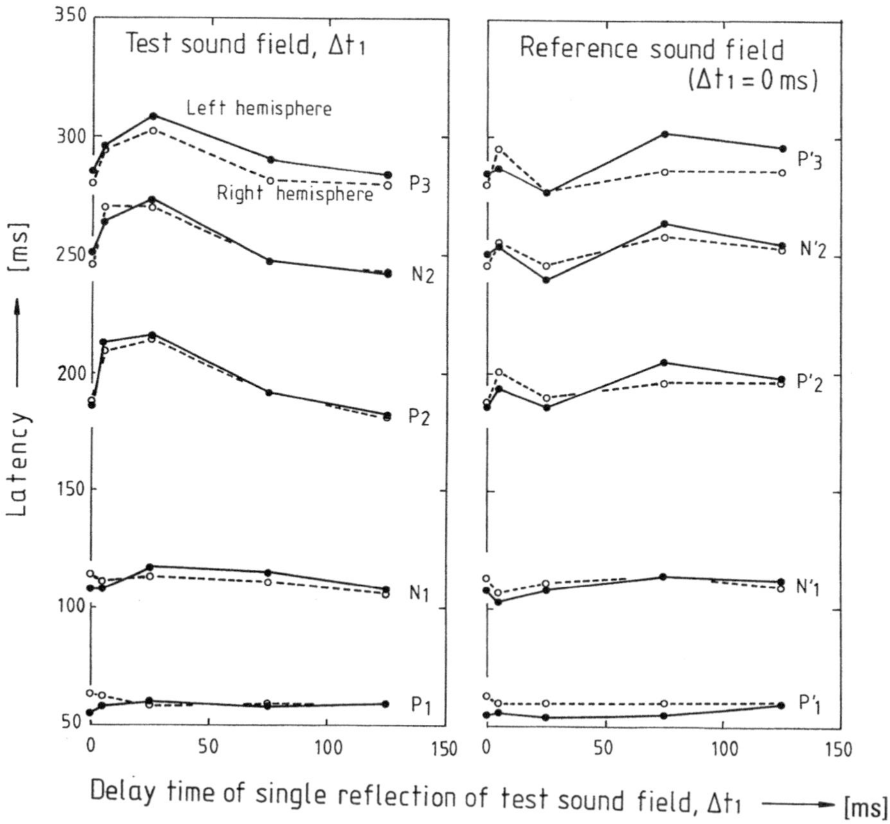

Fig. 2.2 Averaged latencies for both the test sound field and the reference sound field for paired stimuli, as a function of the delay time of the reflection, Δt_1. (____): *left hemisphere*; (------): *light hemisphere*
Maximum latencies of P_2, N_2, and P_3 are found at $\Delta t_1 = 25$ ms for the test sound field, whereas relatively short latencies of P'_2, N'_2, and P'_3 are observed for the reference sound field. This is typical brain activity showing "relativity"

7. Consequently from 40 to 170 ms of the SVR, the hemispheric dominance may be found for the amplitude component, which may be related to respective functional specializations of the hemispheres. Early latency differences corresponding to the SL may be found in the range between 120 and 170 ms.
8. Finally, we found that the N_2-latency components in the delay range between 200 ms and 310 ms may correspond well with subjective preference relative to the listening level, to the time delay of the reflection, and, indirectly, to the IACC.
9. Since the longest latency was always observed for the most preferred condition, one might speculate that the brain is most relaxed at the preferred condition and that this causes the observed latency behavior to occur. Therefore, a correlation

may exist between the duration of latency periods and the duration of alpha wave periods in electroencephalography (EEG) and magnetoencephalography (MEG) during the human waking stage as discussed in the following chapters.

2.2 Electroencephalographic (EEG) and Magnetoencephalographic (MEG) Correlates of Subjective Preference of the Sound Field

In order to attain further knowledge of brain activities, we conducted a series of experiments to find EEG correlates of subjective preference. Since the effects of changes in reverberation time (T_{sub}) could not be seen from auditory-evoked potentials by applying short signals less than 0.9 s in duration, we had to use longer time frames to enable continuous sounds to find some distinctive feature in EEG signals that would follow changes in the T_{sub}. First, we changed the delay time of the single reflection (Δt_1), which reconfirmed the SVR results. Analysis of the data led us to discover an EEG correlate of subjective preference in the effective duration of the autocorrelation function of the alpha wave range (Ando 2009), *i.e., in the duration that alpha rhythms persist in the brain* (Fig. 2.3). *To discriminate*

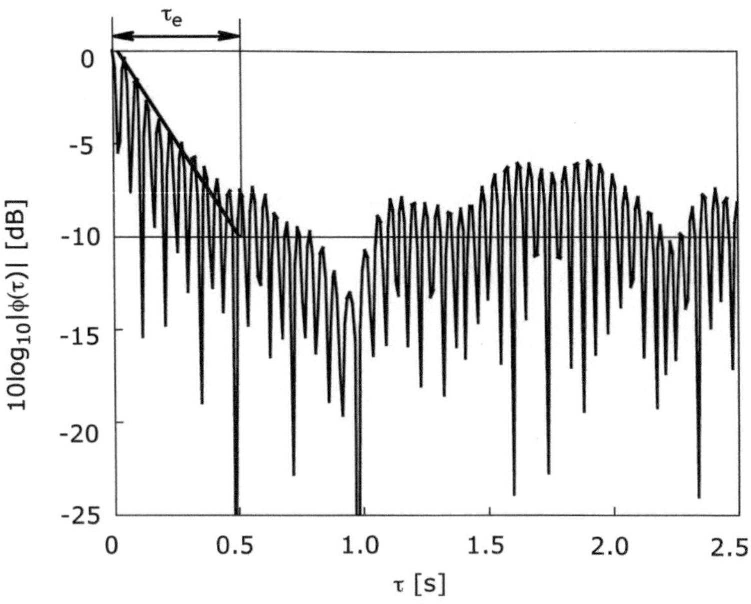

Fig. 2.3 Determining the effective duration (τ_e) of the alpha rhythm by estimating the slope of the envelope of the autocorrelation function and determining the delay at which it reaches 10 % of its maximal at zero-lag value of 0 dB. Effective duration measures duration of temporal coherence, i.e., the duration over which a repetitive structure persists in a signal. Such activity in the alpha rhythm is an indication of subjective preference that may be measured even in newborn babies

more clearly between individual perceptions, we investigated further by using MEG to chart individual responses following changes to Δt_1*. Ultimately, effects of the typical temporal factor (*T_{sub}*) and spatial factor (IACC) were searched in EEG recordings.*

2.2.1 EEG in Response to Change of Δt_1

In this experiment, music motif B (Arnold's Sinfonietta, Opus 48, a 5-s segment of the 3rd movement) was selected as the sound source (Burd 1969; Ando 1985). The delay time of the first, single reflection Δt_1 was alternatively adjusted to 35 ms (a preferred condition) and 245 ms (a condition of clear echo percept). The EEG of ten pairs from positions T3 and T4 was recorded for about 140 s in response to ten delay-change pairs, and experiments were repeated over a total of 3 days. Eleven 22–26-year-old subjects participated in the experiment. Each subject was asked to close their eyes while listening to the music during the recording of the EEG. Two loudspeakers were arranged in front of the subject. Thus, the IACC was kept at a constant value near unity. The sound-pressure level was fixed at a 70 dBA peak, in which the amplitude of the single reflection was the same as that of the direct sound, $A_0 = A_1 = 1$. The leading edge of each sound signal was recorded as a time stamp so that the recorded EEG responses could be precisely synchronized to and correlated with the presented sound. EEG signals amplified, passed through a filter with a 5–40 Hz bandwidth and a slope of 140 dB/octave, and were then digitally sampled at 100 Hz or more.

In order to find brain activity patterns corresponding to subjective preference, we analyzed the effective duration τ_e of the ACFs in the α-wave range (8–13 Hz) of the EEG. First, considering that the subjective preference judgment needs at least 2 s to develop a psychological present, the running integration interval (2T) was examined for periods between 1.0 s and 4.0 s. Remarkable findings are:

1. A satisfactory duration $2T = 2$–3 s in the ACF analysis was found only from the left hemisphere (Ando and Chen 1996). Values of τ_e at $\Delta t_1 = 35$ ms are significantly longer than at $\Delta t_1 = 245$ ms ($p < 0.01$) only in the left hemisphere, but not in the right (Fig. 2.4).
2. Ratios of τ_e values in the α-wave range at $\Delta t_1 = 35$ ms and 245 ms, for each subject, are shown in Fig. 2.5. Remarkably, all individual data indicate that the ratios in the left hemisphere at the preferred condition of 35 ms are much longer than in the right hemisphere.
3. The results reconfirm that when Δt_1 is changed, the left hemisphere is highly activated, and the values of τ_e of α-rhythms in this hemisphere correlate well with subjective preference. The α-rhythm has the longest period in the EEG in the waking state of the adult human and may indicate feelings of "pleasantness" and "comfort" – a preferred condition, which is widely accepted. A long

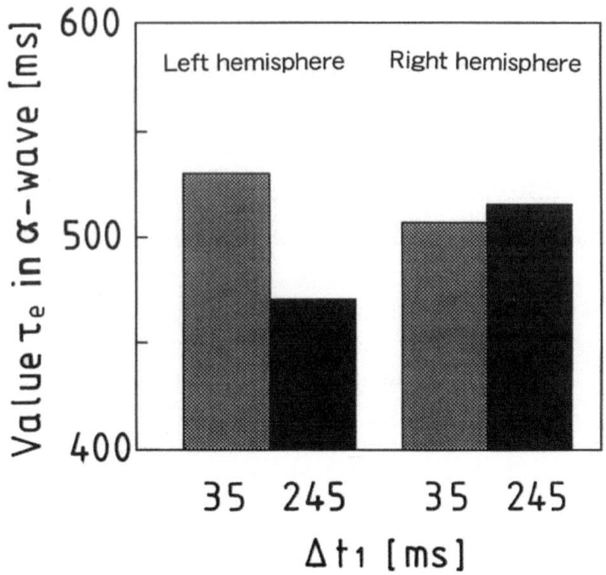

Fig. 2.4 Averaged ACF τ_e values of the EEG alpha wave upon change of Δt_1 between 35 ms and 245 ms (11 Subjects). *Left*: left hemisphere. *Right*: right hemisphere
Significant difference in ACF τ_e values may be found in the left hemisphere associated with the temporal factors, but not in the right hemisphere (associated with spatial factors)

effective duration τ_e of α-rhythms may relate to the long N_2-latency of the SVR in the preferred condition that was shown in Fig. 2.1.

2.2.2 MEG in Response to Change of Δt_1

In MEG studies, the weak magnetic fields produced by electric currents flowing in neurons are measured with a multiple channel SQUID (superconducting quantum interference device for magnetism detection) gradiometers, which enables the study of many interesting properties of the working human brain. MEG accurately detects superficial tangential currents, whereas EEG is sensitive to both radial and tangential current sources and reflects activity in the deepest parts of the brain. Only currents that have a component tangential to the surface of a spherically symmetric conductor produce a sufficiently strong magnetic field outside of the brain; radial sources are thus externally silent. Therefore, MEG mainly measures neuronal activity from the fissures of the cortex, which often simplifies interpretation of the data. Fortunately, all the primary sensory areas of the brain – auditory, somatosensory, and visual – are located within fissures. The advantages of MEG over EEG result mainly from the fact that the skull and other extracerebral tissues are practically transparent to magnetic fields but do substantially alter electrical

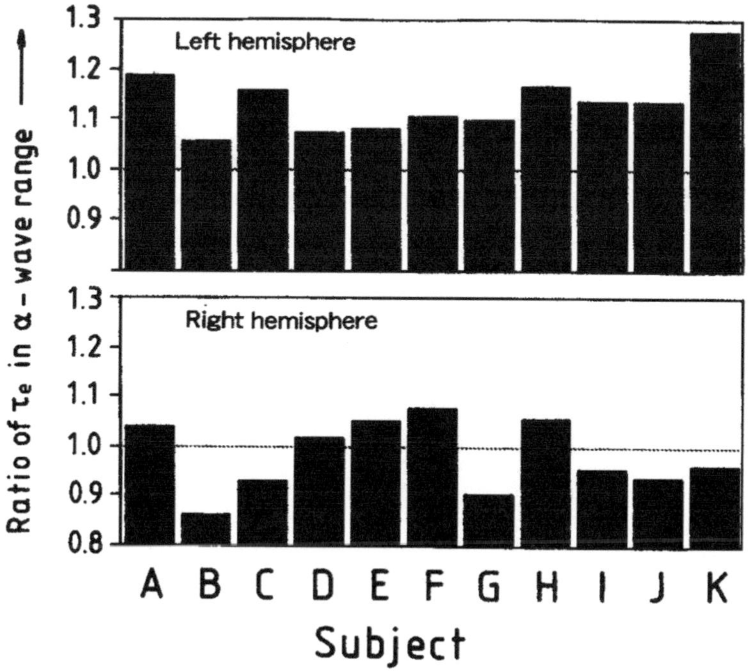

Fig. 2.5 Ratios of ACF τ_e values of the EEG alpha wave upon change of Δt_1 between 35 ms and 245 ms for each individual subject, A–K: [τ_e value at 35 ms]/[τ_e value at 245 ms]
Above: left hemisphere. *Below*: right hemisphere
Ratios of ACF τ_e values of the EEG alpha wave are always greater on the left hemisphere than on the right

current flows. Thus, magnetic patterns outside the head are less distorted than the electrical potentials on the scalp. Further, magnetic recording is reference-free, whereas electric brain maps depend on the location of the reference electrode.

Measurements of MEG responses were performed in a magnetically shielded room using a 122-channel whole-head neuromagnetometer as shown in Photo 2.1 (Neuromag-122^{TM}, Neuromag Ltd., Finland) (Soeta et al. 2002

The source signal was the word "piano," which had a 0.35-s duration. The minimum value of the moving effective duration τ_e, i.e., $(\tau_e)_{min}$, was about 20 ms. It is worth noting that this value is close to the most preferred delay time of the first reflection of sound fields with continuous speech (Ando and Kageyama 1977). In the present experiment, the delay time of the single reflection (Δt_1) was set at five levels (0, 5, 20, 60, and 100 ms). The direct sound and a single reflection were mixed, and the amplitude of the reflection was the same as that of the direct sound ($A_0 = A_1 = 1$). The auditory stimuli were binaurally delivered through metal-free plastic tubes and earpieces into the ear canals. The sound-pressure level, which was measured at the end of the tubes, was fixed at 70 dBA.

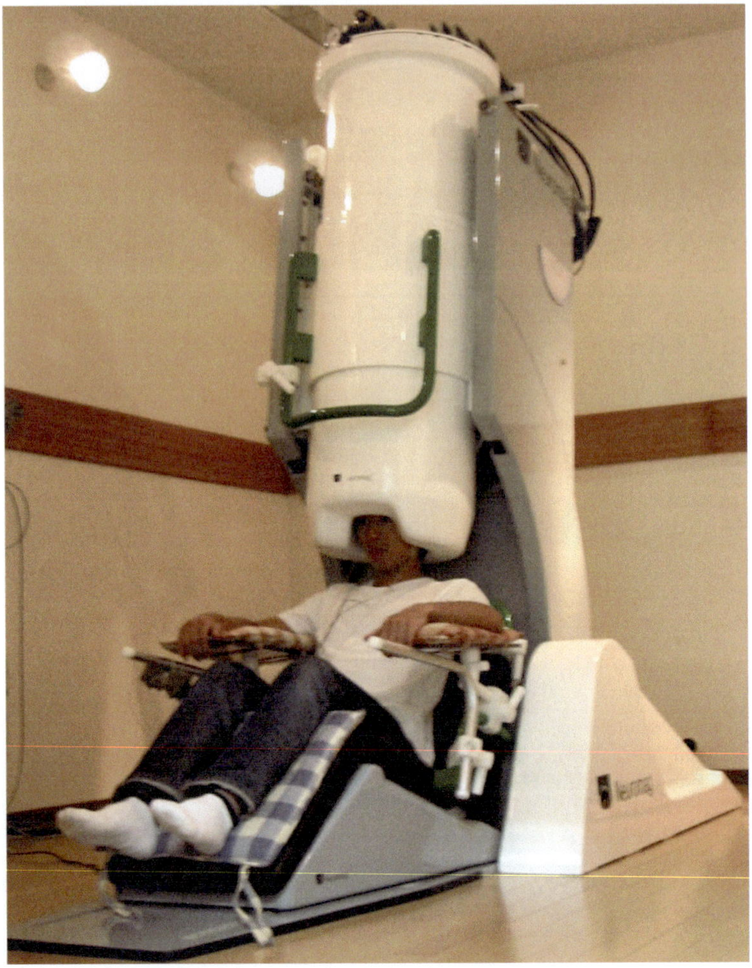

Photo 2.1 Magnetometer used in recording the magnetoencephalogram (MEG)

Seven 23–25-year-old subjects with normal hearing participated in the experiment. In accordance with the PCT methodology, each subject compared ten possible stimulus pairs per session, and a total of ten sessions were conducted for each subject. Measurements of magnetic responses were performed in a magnetically shielded room. Similar to the EEG measurements described in the previous section, the paired-auditory stimuli were presented in the same way as in the subjective preference test. During measurements, the subjects sat in a chair with their eyes closed. To compare the results of the MEG measurements with the scale values of the subjective preference, combinations of a reference stimulus ($\Delta t_1 = 0$ ms) and test stimuli ($\Delta t_1 = 0, 5, 20, 60$, and 100 ms) were presented alternately 50 times, and the MEG signals were analyzed. The magnetic data were recorded continuously with a filter of 0.1–30.0 Hz and digitized with a sampling rate of 100 Hz. Eight channels that had larger amplitudes of N1m response in each hemisphere were selected for the ACF analysis. We analyzed the MEG alpha band signal for each of the paired stimuli for each subject, computed their normalized autocorrelation functions, and plotted them using logarithmic units of magnitude. The envelope of the ACF is fitted with a straight line and its intercepts with the –5 and –10 dB magnitude values are determined. The –10 dB (10 % of maximum value) intercept is defined as the effective duration (Fig. 4.27 in Ando 2009). Naturally, for the preferred condition at $\Delta t_1 = 5$ ms where the first reflection time is short, the computed effective duration value of the MEG alpha band response is long ($\tau_e \approx 0.5$ s), compared to the effective duration of $\tau_e \approx 0.3$ s that is seen for the unfavorable condition of echo disturbance ($\Delta t_1 = 100$ ms).

The results from the eight subjects confirmed a linear relationship between the averaged τ_e values in the MEG alpha band and the averaged scale values of subjective preference. The left hemisphere dominates the temporal factor Δt_1, which reconfirms the aforementioned SVR and EEG studies. Therefore, we analyzed individual results from the left hemispheres of the subjects. We concluded that:

1. An almost direct relationship between individual scale values of subjective preference and the τ_e values over the left hemisphere was found in each of the eight subjects. Results for each of the eight subjects are shown in Fig. 2.10.
2. Remarkably, the correlation coefficient, r, was above 0.94 for all subjects.
3. It is worth noting that there was only weak correlation between the scale values of subjective preference and the amplitude of the α-wave, $\Phi(0)$, for either hemisphere ($r < 0.37$).
4. The value of τ_e in the alpha wave band is the persistence of alpha rhythms in time, so that under the preferred conditions, the brain repeats a similar rhythm of alpha activity for a longer period of time. This tendency for a longer effective duration τ_e of the alpha rhythm under the preferred condition is much more significant than the aforementioned results that were obtained through similar analyses of EEG signals.

2.2.3 EEG in Response to Change of T_{sub}

Now, let us examine how effective durations τ_e of α-rhythms change in response to subsequent reverberation times (T_{sub}) and their subjective preference values. Ten student subjects participated in the experiment (Chen and Ando 1996). The sound source used was music motif B, herein described as having a minimum effective duration of $(\tau_e)_{min} \sim 40$ ms, so that the most preferred reverberation time can be calculated as $(T_{sub})_p \sim 23\ (\tau_e)_{min} = 0.92$ s (Eq. 6.5). Ten 25–33-year-old subjects participated in the experiment. EEG signals from the left and right hemispheres were recorded. Values of τ_e in EEG signals in the α-band were also analyzed for the duration of $2T = 2.5$ s.

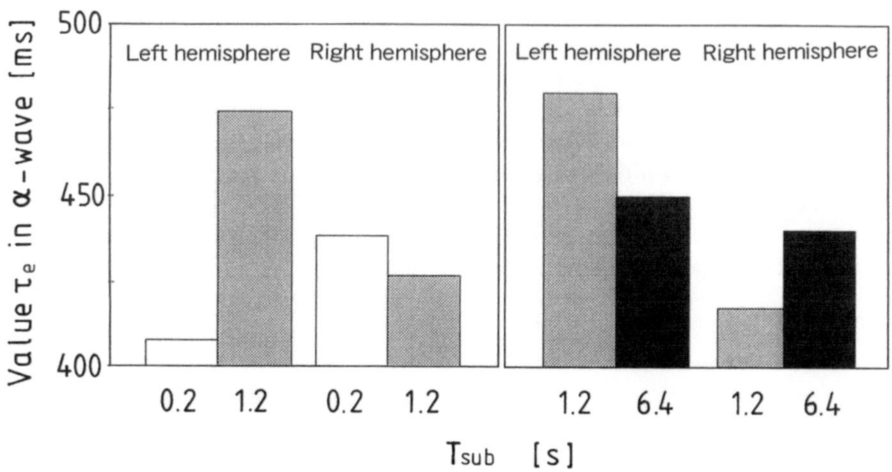

Fig. 2.6 Averaged value of ACF τ_e of the EEG alpha wave upon change of T_{sub}: 0.2 s and 1.2 s; 1.2 s and 6.4 s for 10 subjects. The *left column* in each figure depicts results from the left hemisphere, and the *right column* depicts results from the right hemisphere

First, let us consider the averaged values of the effective duration τ_e of α-activity, shown in Fig. 2.6.

1. Clearly, for the left hemisphere the values of τ_e are much longer close to the preferred condition 0.92 s, i.e., at $T_{sub} = 1.2$ s, than at either of the non-preferred conditions $T_{sub} = 0.2$ s and 6.4 s.
2. Thus, alpha rhythms persist longer in the left hemisphere for preferred reverberation time values. However, the contrary is true for the right hemisphere, where the non-preferred reverberation times produced longer effective durations of alpha activity.

The results of the analysis of variance (ANOVA) reveal that although there are vast individual differences, a significant difference is achieved for T_{sub} in the pair of 0.2 s and 1.2 s ($p < 0.05$), and interference effects are observed for factors Subject and LR ($p < 0.01$), and LR and T_{sub} ($p < 0.01$). No such significant differences are achieved for the pair at 1.2 s and 6.4 s, but there are interference effects between subject and LR, and subject and T_{sub}. Different individual scale values of preference, however, are well correlated to ratios of values of τ_e for the α-wave band in the left hemisphere (Chen and Ando 1996; Ando 1998, 2009).

2.2.4 EEG in Response to Change of the IACC

The EEG response to changes in the IACC was also investigated. Eight student subjects participated in the paired-comparison experiment (Sato et al. 2003). Music

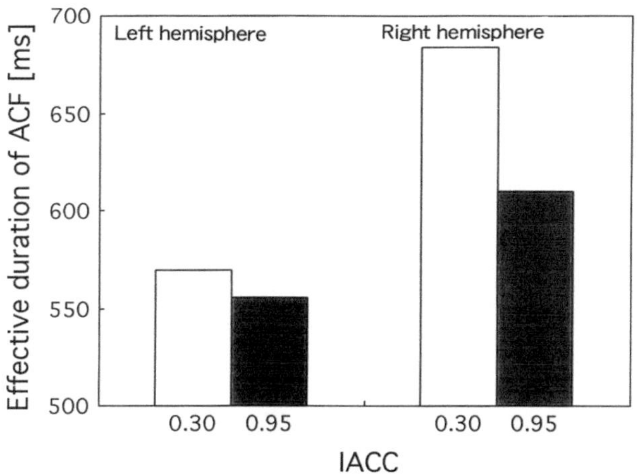

Fig. 2.7 Averaged values of ACF τ_e of the EEG alpha wave upon change of the IACC for the pair of IACC = 0.30 and 0.95. The *left column* in the figure is from the left hemisphere, and the *right column* is from the right hemisphere

motif B was again used as the stimulus. Changes in the IACC reflected clearly in right hemisphere dominance. The effective duration τ_e of α-band activity was found to be substantially longer in the preferred condition (IACC = 0.30). A significant difference was achieved in the right hemisphere for the pair of sound fields with IACC = 0.95 and 0.30 ($p < 0.01$) as shown in Fig. 2.7.

1. In seven of eight subjects, the ratios of effective durations τ_e for α-band responses to IACC change, [τ_e (IACC = 0.3)/τ_e (IACC = 0.95)], in the right hemisphere were greater than in the left hemisphere except for subject B (Fig. 2.8). Thus, as far as the IACC is concerned, the more preferred condition with a smaller IACC is related to longer α-rhythm effective durations in the right hemisphere in most of subjects tested.
2. As shown the most clearly in Fig. 2.9, a wave of alpha rhythm activity in the right hemisphere (T4) at IACC = 0.3, later propagates toward the left hemisphere (T3).
3. Additionally, experiments using MEG measurements tested a speech signal that was altered by changes made to the IACC (0.27, 0.61 and 0.90). The results reconfirmed that effective duration τ_e and the maximum amplitude of the CCF increased when the IACC decreased in the right hemisphere (Soeta et al. 2005).

Table 2.1 summarizes hemispheric dominance results obtained by analysis of the effective durations τ_e of α-rhythms, with respect to changes in listening level LL, first reflection time Δt_1, reverberation time T_{sub}, and the interaural crosscorrelation coefficient IACC. This finding suggests that the value of τ_e in the α-band is an objective index for designing excellent human acoustic environments.

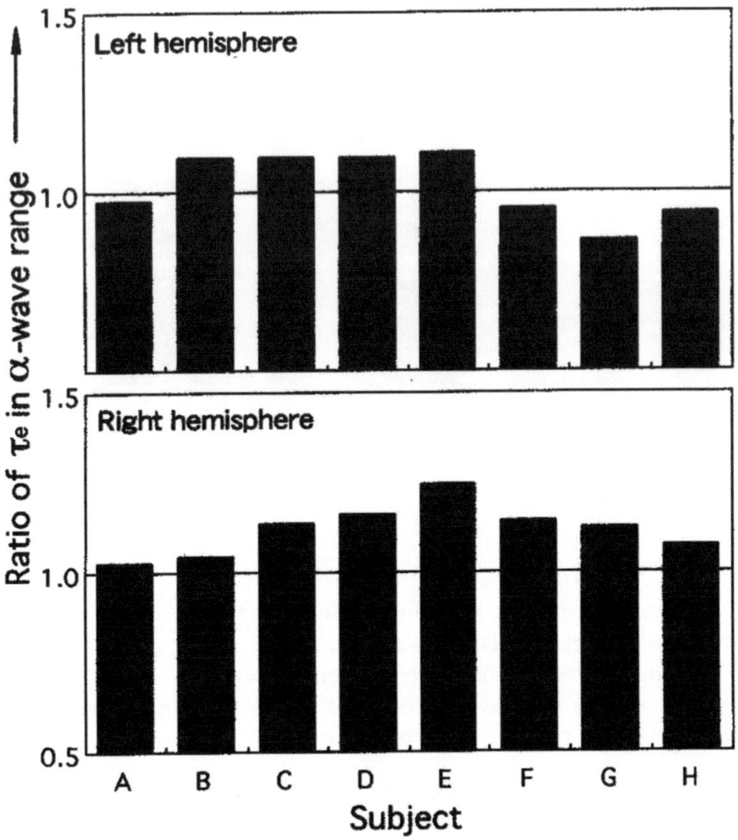

Fig. 2.8 Ratio of ACF τ_e values of the EEG alpha wave from the left hemisphere (T3) and the right hemisphere (T4) for each of 8 subjects, A–H: [τ_e value at IACC = 0.30]/[τ_e value at IACC = 0.95]. Ratio of ACF τ_e values is greater on the right hemisphere than on the left except for subject B

2.3 Specialization of Cerebral Hemispheres for the Sound Field

Four significant, orthogonal physical factors for describing sound fields have been found. Two of them are temporal aspects, i.e., the initial time delay gap between the direct sound and the first reflection, Δt_1, and the subsequent reverberation time, T_{sub}, while the other two are spatial aspects, i.e., the magnitude of the interaural crosscorrelation function, the IACC, and the volume of the binaural listening level, LL. Specialization of the cerebral hemispheres according to the temporal and spatial factors when each of these four factors is changed during the experiments has been identified.

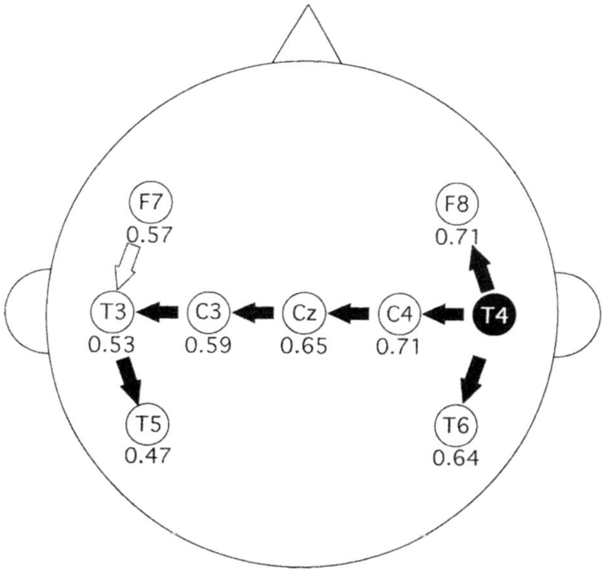

Fig. 2.9 Propagation of the alpha wave flow from the right hemisphere to the left in response to a change of IACC. Real numbers reflect the median values of alpha rhythm correlation magnitudes (maximum absolute values) between alpha band EEG signals from electrode T4 and the indicated electrodes

Table 2.1 Hemispheric specializations determined by AEP, EEG, and MEG of the left and right hemispheres for temporal and spatial factors of the sound field, respectively

Factors changed	AEP (SVR) $A(P_1 - N_1)$	EEG, ratio of ACF τ_e values of α-wave	AEP (MEG) N1m	MEG, ACF τ_e value of α-wave
Temporal				
Δt_1	$L > R$ (speech)[1]	$L > R$ (music)		$L > R$ (speech)
T_{sub}	–	$L > R$ (music)	–	
Spatial				
LL	$R > L$ (speech)	–	–	
IACC	$R > L$ (vowel/ *a/*)	$R > L$ (music)[2]	$R > L$ (band noise)[3]	
	$R > L$ (band noise)			
τ_{IACC}			$R > L$ (band noise)[3]	
Head-related transfer functions			$R > L$ (vowels)[4]	

[1]Sound source used in experiments is indicated in the bracket
[2]The flow of EEG α-wave from the right hemisphere to the left hemisphere in response to music stimulus in change of the IACC was determined by the CCF $|\phi(\tau)|_{max}$ between α-waves recorded at different electrodes
[3]Soeta and Nakagawa (2006)
[4]Palomaki et al. (2002)

Recordings of brain activity over the left and right hemispheres, including the slow-vertex response (SVR), electroencephalogram (EEG), and magnetoencephalogram (MEG), have revealed evidences as listed in Table 2.1. Formulation of such a neurally grounded strategy for acoustic design has been initiated through a study of auditory-evoked electrical potentials, i.e., the slow-vertex responses (SVR), which are generated by the left and right human cerebral hemispheres. The goal of these experiments was to identify potential neuronal response correlates of subjective preference for the orthogonal acoustic parameters related to sound fields. Using the paired-comparison method, it had been established that particular ranges of the four orthogonal factors were preferred by most listeners. These factors and auditory-evoked potentials (AEPs) are integrated by the triggering technique, so that reliable predictions of subjective preference could be made. Similarly, in the study the SVR for paired stimuli was integrated, and scale values of subjective preference based on the paired-comparison method were obtained. Typically, SVRs are the strongest when stimulus patterns change abruptly, i.e., there is a contrast between the paired stimuli in which spatial and/or temporal factors change. The method of paired stimuli is therefore the most effective procedure because of this relativity of the brain response.

Main results are as follows:

1. The left- and right-relative peak amplitudes of the first major SVR waves indicate lateralization of responses associated with the temporal factor (Δt_1 in the left hemisphere) and spatial factors (LL and IACC in the right hemisphere) (Fig. 2.1; Ando et al. 1987a, b; Ando 2003). Correlates of loudness, the sensation level SL and the binaural listening level LL, were classified as temporal-monaural factors from a physical viewpoint. However, results of the SVR indicate that the neural correlates of these parameters were predominantly observed over the right hemisphere. Thus, SL and LL should be reclassified as spatial factors as far as specialization of the hemispheres is concerned. Classification of the LL as a spatial factor is natural because it is accurately measured by the geometric average of sound energies arriving at the two ears.
2. Results of EEG recordings that showed hemispheric lateralization of the temporal factors, i.e., Δt_1 and T_{sub} (reverberation time), reconfirmed left hemisphere dominance for these factors (Ando and Chen 1996; Chen and Ando 1996). Similarly, spatial factors associated with the interaural magnitude IACC showed right hemispheric dominance (Sato et al. 2003; Sato and Prodi 2009). It is worth noting that the scale value of subjective preference has a tendency to increase as IACC decreases. Thus, judging by the evidence provided by the measured subjective neural responses to varying temporal and spatial factors, there appears to be a high degree of independence between the left and right hemispheres.
3. Scale values of subjective preferences are well predicted from the values of the effective durations τ_e (extracted from the ACF) of EEG alpha band activity (8–13 Hz) over the left and right hemispheres. Neural correlates of preferences related to changing temporal and spatial factors of the sound fields dominate in the left and right hemisphere EEG signals, respectively (Figs. 2.7 and 2.8).

4. Recorded MEG amplitudes reconfirmed the left hemisphere specialization for the first reflection time Δt_1 (Soeta et al. 2002).
5. Scale values of individual subjective preferences relate directly to the value of effective duration τ_e extracted from the ACF of the alpha band activity of the MEG (Fig. 2.10). Note that it is the effective durations of the alpha rhythms in EEG and MEG recordings, and not the absolute amplitudes of these waves, that correspond to scale values of subjective preferences.
6. In addition to the temporal response patterns described above, propagation of a wave of alpha rhythm activity in the right hemisphere from electrode T4 to the whole brain including the left hemisphere is demonstrated in Fig. 2.9. Here, spatial patterns of cortical neural response were analyzed by examining crosscorrelations between alpha band activity at different scalp locations in the two hemispheres using EEG and MEG recordings. The results indicate alpha rhythm activity over larger areas of the cerebral cortex in response to preferred sound fields (Soeta et al. 2003). These findings indicate that the brain repeats a similar temporal rhythm in the alpha frequency range over a wider area of the scalp when preferences are better satisfied. It is also noteworthy that the alpha rhythm has the longest period of brain waves in the waking state and is deeply related to relaxation and health. Previously, the left hemisphere has been mainly associated with identification of time sequences and functions, and the right hemisphere has been known to be fundamentally concerned with spatial identification.

Left hemispheric specialization for speech signals has been reported by a number of authors using EEG and MEG recordings. However, when the IACC was changed for speech and music signals, right hemispheric dominance was observed as indicated in Table 2.1. Therefore, rather than having an absolute response bias for particular kinds of signals, hemispheric response dominance is relative and depends on which factor is changed in the comparison pair.

These findings and the theory of subjective preference reconfirm the central auditory signal-processing model (Ando 1985) as shown in Fig. 2.11 (Ando 1998, 2007, 2009).

2.4 Model of Auditory Brain System

Based on the above-mentioned physical system and physiological responses, a high-level signal-processing model of the central auditory system (Ando 1985) *has been reconfirmed as shown in* Fig. 2.11. *Applying this model can yield a wide range of research works. For example, research has been initiated for enabling automatic speech recognition over the telephone and in rooms, which would allow identifying which one of the 7000 languages that are alive in the world today is being spoken* (Ando 2015). *If the neural ACF is extended in time to cover longer time delays than those present in neural responses, then a running neural ACF*

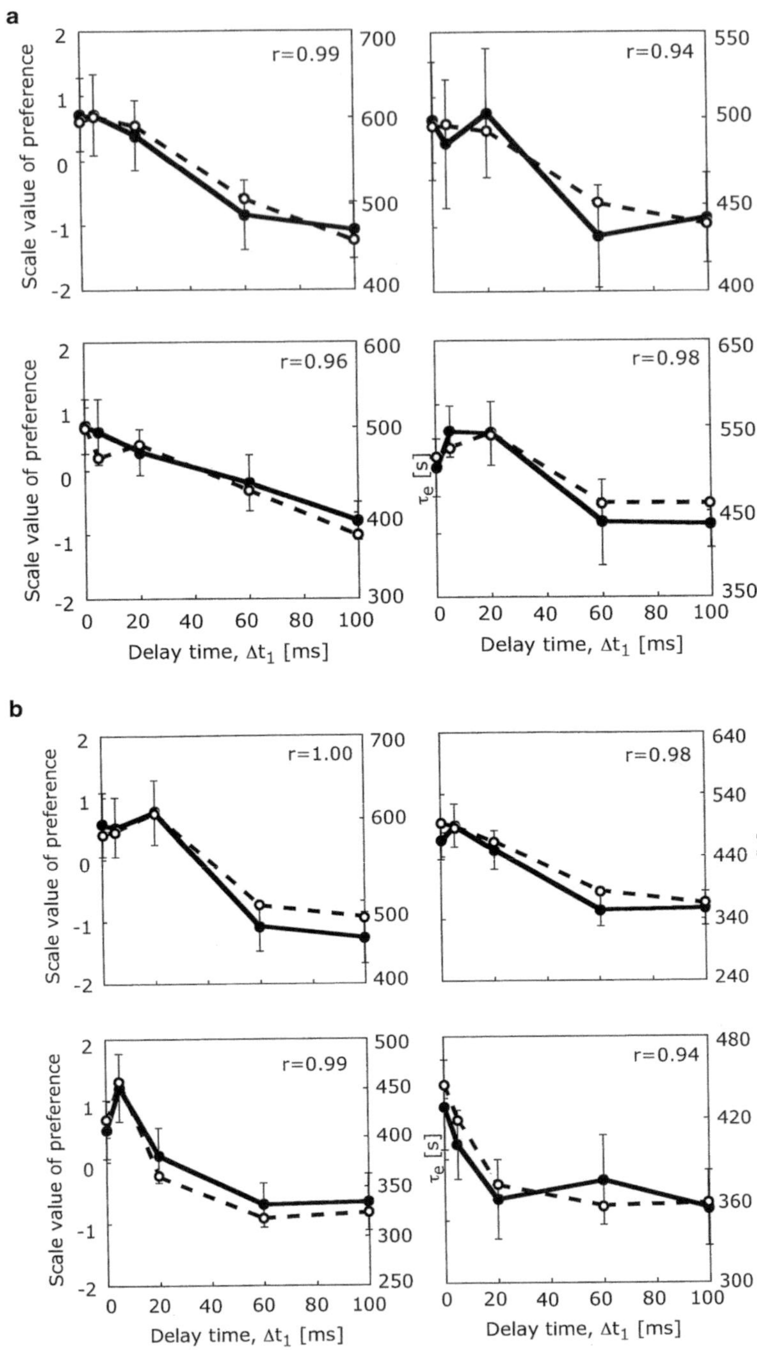

Fig. 2.10 Good correspondence between the scale value of subjective preference and the averaged ACF τ_e value of the MEG alpha wave over the left hemisphere (8 subjects). The averaged τ_e value and the scale value are the highest correlation over the eight channels. ○: scale values of subjective preference. ●: averaged τ_e values of MEG alpha wave, error bars being standard errors

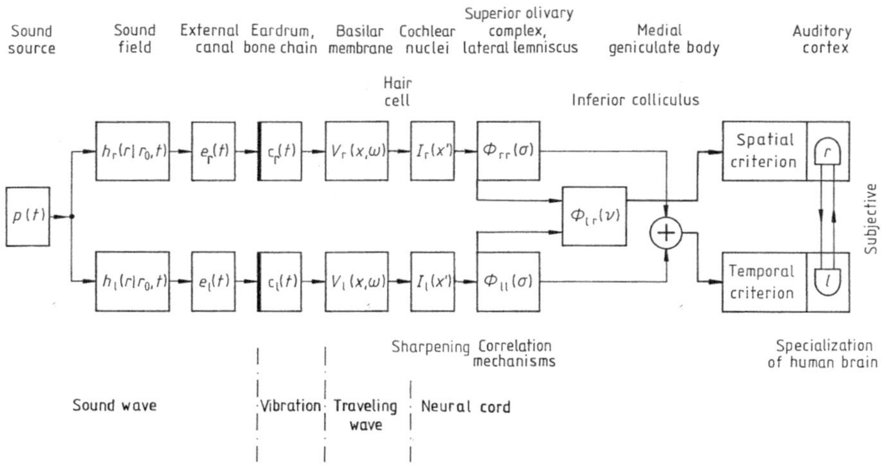

Fig. 2.11 Central auditory signal-processing model for subjective responses (Ando 1985) $p(t)$: source sound signal in the time domain. $h_{l,r}(r/r_0,t)$: head-related impulse responses from a source position of r_0 to the left and right ear entrances of a listener at r. $e_{l,r}(t)$: impulse responses of left and right external canals, from the left and right ear entrances to the left and right eardrums. $c_{l,r}(t)$: impulse responses for vibration of left and right bone chains, from the eardrums to oval windows, including transformation factors into vibration motion at the eardrums – $V_{l,r}(x,\omega)$: travelling wave forms on the basilar membranes, where x is the position along the left and right basilar membrane measured from the oval window. $I_{l,r}(x')$: sharpening in the cochlear nuclei corresponding to roughly the power spectra of input sound, i.e., responses of a single pure tone ω tend to a limited region of nuclei. These neural activities may be enough to be converted into activities similar to the ACF. $\Phi_{ll}(\sigma)$ and $\Phi_{rr}(\sigma)$: ACF mechanisms in the left and right auditory pathways, respectively. Symbol $\oplus$ signifies that signals are combined. $\Phi_{lr}(\nu)$: IACF mechanism (Ando 1985). r and l: specialization for temporal and spatial factors of the left and right human cerebral hemispheres, respectively. Temporal and spatial percepts (Sect. 4.2) may be processed in the left and right hemisphere according to the temporal factors extracted from the ACF and the spatial factors extracted from the IACF, respectively. The overall subjective preference and annoyance may be processed in both hemispheres in relation to the temporal and spatial factors (Ando 2002, 2009)

could potentially come to aid in a broader spectrum of societal activities. It would be possible to classify music based on the effective duration of the ACF $(\tau_e)_{min}$ *of musical notes in order to select concert programs that blend preferred temporal factors with ideal acoustic spaces* (Ando 2009)*; use automatic identification of noise source by implementing ACF factors for the purposes of reducing aircraft or traffic noise* (Soeta and Ando 2015)*; study the effects that environmental noise has on the developing brain and its specialization and the accumulated effects that noise has on the mind* (Ando 2011).

The model consists of monaural autocorrelation mechanisms, the interaural crosscorrelation mechanism between the two auditory pathways, and the specialization of the human cerebral hemispheres for temporal and spatial factors of the sound field. In addition, according to the relationship between temporal and spatial percepts of sound as well as subjective preference of the sound field and changes in physiological phenomena in response to variations in the acoustic factors, a model is reconfirmed and formed as shown in Fig. 2.11 (Ando 1998, 2009). In this figure, a sound source p(t) is located at position r_0 in a three-dimensional space, and a listener is sitting at position r, which is defined by the location of the center of the head. $h_{l,r}(r| r_0,t)$ are the impulse responses of the two sound paths through the room between r_0 and the left and right ear canal entrances. The impulse responses of the external ear canal and the bone chain are $e_{l,r}(t)$ and $c_{l,r}(t)$, respectively. The velocity of the basilar membrane is expressed by $V_{l,r}(x, \omega)$, x being the position along the membrane.

In fact, the time domain analysis, rather than the frequency domain analysis, of firing rates of cat auditory nerve fibers revealed a pattern of the ACF (Secker-Walker and Searle 1990).

In contrast with neural representations that are based on neural firing rates, which tend to change their form and degrade at higher sound levels, representations based on temporal spike information remain largely invariant in form over the whole dynamic range of hearing and thus mirror the behavior of auditory percepts. For stimuli consisting of low-frequency components that are resolved by cochlear filters, "pooled all-order interspike interval distributions" or simply "population-interval distributions" resemble the autocorrelation of the stimulus itself. For sound stimuli consisting of higher-frequency components that are not resolved by cochlear filters, population-interval distributions resemble the autocorrelation of their waveform envelopes (Cariani and Delgutte 1996). From a viewpoint of the missing fundamental or pitch of complex components judged by human perception, the running ACF must be processed in the frequency components below about 5 kHz and the fundamental frequency below 1200 Hz (Inoue et al. 2001).

As we have discussed, evidence shows that there exist correlates of the interaural correlation magnitude IACC in neural activity at the level of the brainstem and midbrain (Ando 1998, 2009). The interaural crosscorrelation mechanism exists at the level of the superior olive and the inferior colliculus. It is concluded that the output signal of the interaural crosscorrelation mechanism that computes the IACC is predominantly connected to the right hemisphere. Representation of sound-pressure level (SPL) may also be preferentially processed in the right hemisphere. Sound-pressure level can be expressed in terms of a geometrical average of the ACFs for the two ears at the origin of time ($\sigma = 0$), and changes in sound-pressure level produce changes in neuronal response latencies that appear at the level of the inferior colliculus.

Based on the model, temporal and spatial percepts (Ando et al. 1999; Ando 2002, 2009) and subjective attributes of sound fields in terms of processes occurring in the auditory pathways as well as the specialization of the two cerebral hemispheres will be described.

References

Ando Y (1977) Subjective preference in relation to objective parameters of music sound fields with a single echo. J Acoust Soc Am 62:1436–1441

Ando Y (1983) Calculation of subjective preference at each seat in a concert hall. J Acoust Soc Am 74:873–887

Ando Y (1985) Concert hall acoustics. Springer, Heidelberg

Ando Y (1992) Evoked potentials relating to the subjective preference of sound fields. Acustica 76:292–296

Ando Y (1998) Architectural acoustics, blending sound sources, sound fields, and listeners. AIP Press/Springer, New York

Ando Y (2002) Correlation factors describing primary and spatial sensations of sound fields. J Sound Vib 258:405–417

Ando Y (2003) Investigations on cerebral hemisphere activities related to subjective preference of the sound field, published for 1983–2003. J Temporal Des Architect Environ 3:2–27

Ando Y (2007) Concert hall acoustics based on subjective preference theory. In: Thomas R (ed) Springer handbook of acoustics. Springer, New York, Chapter 10

Ando Y (2009) Auditory and visual sensations. Springer, New York

Ando Y (2011) Brain oriented acoustis. Ittosha, Tokyo (in Japanese)

Ando Y (2015) Autocorrelation-based features for speech representation. Acta Acustica United Acustica 101(2015):145–154

Ando Y, Chen C (1996) On the analysis of the autocorrelation function of α-waves on the left and right cerebral hemispheres in relation to the delay time of single sound reflection. J Journal of Architectural Planning and Environmental Engineering. J Architect Plann Environ Eng Architect Inst Jpn (AIJ) 488:67–73

Ando Y, Kageyama K (1977) Subjective preference of sound with a single early reflection. Acustica 37:111–117

Ando Y, Kang SH, Nagamatsu H (1987a) On the auditory-evoked potentials in relation to the IACC of sound field. J Acoust Soc Jpn (E) 8:183–190

Ando Y, Kang SH, Morita K (1987b) On the relationship between auditory-evoked potential and subjective preference for sound field. J Acoust Soc Jpn (E) 8:197–204

Ando Y, Sato S, Sakai H (1999) Fundamental subjective attributes of sound fields based on the model of auditory-brain system. In: Sendra JJ (ed) Computational acoustics in architecture. WIT Press, Southampton

Botte MC, Bujas Z, Chocholle R (1975) Comparison between the growth of averaged electroencepharic response and direct loudness estimations. J Acoust Soc Am 58:208–213

Burd AN (1969) Nachhallfreir Musik fuer akustische Modelluntersuchungen. Rundfunktechn Mitteilungen 13:200–201

Cariani PA, Delgutte B (1996) Neural correlates of the pitch of complex tones. I. Pitch and pitch salience. J Neurophysiol 76:1698–1716

Chen C, Ando Y (1996) On the relationship between the autocorrelation function of the α-waves on the left and right cerebral hemispheres and subjective preference for the reverberation time of music sound field. J Architect Plann Environ Eng Architect Inst Jpn (AIJ) 489:73–80

Inoue M, Ando Y, Taguti T (2001) The frequency range applicable to pitch identification based upon the auto-correlation function model. J Sound Vib 241:105–116

Palomaki K, Tiitinen H, Makinen V, May P, Alku P (2002) Cortical processing of speech sounds and their analogues in a spatial auditory environment. Cogn Brain Res 14:294–299

Sato S, Nishio K, Ando Y (2003) Propagation of alpha waves corresponding to subjective preference from the right hemisphere to the left with change in the IACC of a sound field. J Temporal Des Architect Environ 3:60–69

Sato S, Prodi N (2009) On the subjective evaluation of the perceived balance between a singer and a piano inside different theatres. Acta Acustica United with Acustica 95:519–526

Secker-Walker HE, Searle CL (1990) Time domain analysis of auditory-nerve- fiber firing rates. J Acoust Soc Am 88:1427–1436

Soeta Y, Ando Y (2015) Neurally based measurement and evaluation of environmental noise. Springer, Tokyo

Soeta Y, Nakagawa S (2006) Auditory evoked magnetic fields in relation to interaural time delay and interaural crosscorrelation. Hear Res 220:106–115

Soeta Y, Nakagawa S, Tonoike M, Ando Y (2002) Magnetoencephalographic responses corresponding to individual subjective preference of sound fields. J Sound Vib 258:419–428

Soeta Y, Nakagawa S, Tonoike M, Ando Y (2003) Spatial analysis of magnetoencephalographic activities in relation to subjective preference for a sound field. J Temporal Des Archit Environ 3:28–35. http://www.jtdweb.org/journal/

Soeta Y, Nakagawa S, Tonoike M (2005) Magnetoencephalographic activities related to the magnitude of the interaural cross-correlation function (IACC) of sound fields. J Temporal Des Archit Environ 5:5–11. http://www.jtdweb.org/journal/

Chapter 3
EEG and MEG Correlates of Visual Subjective Preferences

3.1 EEG Correlates of Subjective Preferences for Oscillatory Horizontal Movements of a Single Target

This study focused on the EEG correlates of subjective preference for oscillatory horizontal movements of a single target when the period of the movement was changed. The EEG was recorded during the presentation of the most preferred and less preferred moving stimuli. The effective duration τ_e in the ACF of the alpha wave was analyzed. Results show that the value of τ_e at the most preferred condition was longer than for the stimulus in the less preferred conditions. In addition, the maximum value of the CCF ($|\phi(\tau)|_{max}$) between EEGs recorded at different electrode sites was analyzed as shown in Fig. 3.1. Results indicate that the value of $|\phi(\tau)|_{max}$ of the alpha wave at the most preferred condition was greater than for the stimulus in the less preferred conditions. These results reconfirm that at the preferred condition, the brain repeats the rhythm in the alpha wave range in the time domain and that this activity spreads over a wider area of the human cerebral cortex.

The aim of this study was to identify the relationship between human brain response and subjective preference of a single circular target moving sinusoidally in the horizontal direction, varying its period (Okamoto et al. 2003). The EEG was recorded during the presentation of stimuli having the most preferred and less preferred periods. Then, the effective durations τ_e of EEG alpha rhythms and their correlated behavior at different electrode sites $|\phi(\tau)|_{max}$ were analyzed to determine the relationship with subjective preference.

Stimuli consisting of white disks with a visual angle of 1.0° in diameter against a black background (Fig. 3.2) were presented on the display placed in front of the subject at a viewing distance of 1.0 m in a dark chamber. The amplitude was fixed at the visual angle of 0.5°. Whereas the most preferred period of the stimuli $[T]_p$ was

Y. Ando, *Brain-Grounded Theory of Temporal and Spatial Design*,
DOI 10.1007/978-4-431-55891-0_3

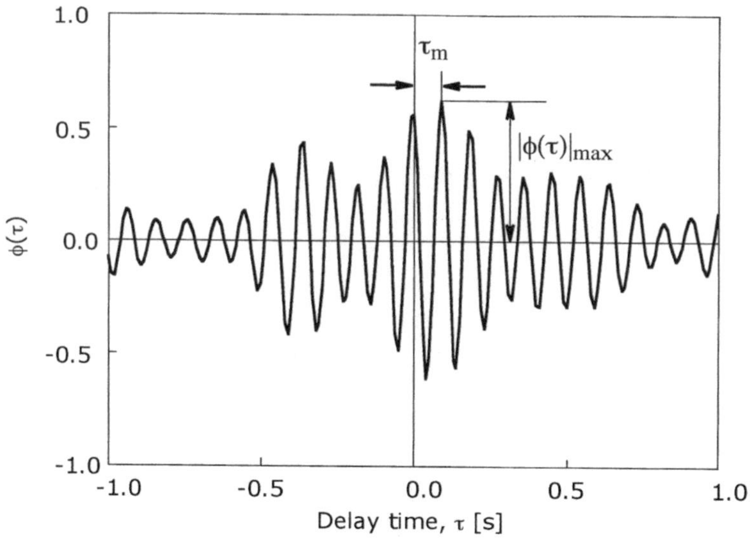

Fig. 3.1 An example of a normalized crosscorrelation function (CCF) between two EEG alpha band signals recorded from different electrodes, showing definition of the maximum correlation value $|\phi(\tau)|_{max}$ and its delay time τ_m

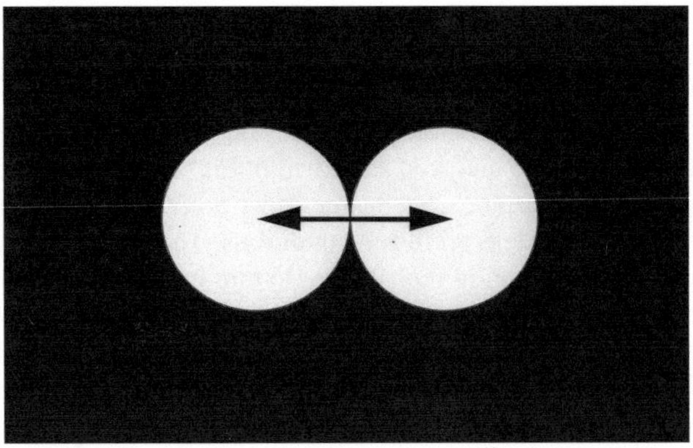

Fig. 3.2 Stimulus target used in the experiment showing an example of oscillatory horizontal movement

found by the PCT at approximately 1.10 s, the subjective preference ratings decreased for shorter and longer periods (Fig. 3.3).

In order to clarify the effect of subjective preference on EEG, stimuli with the most preferred and less preferred periods were selected as paired stimuli. The thick line shows the preference evaluation curve obtained from the results above. Two pairs of stimuli were presented to determine if either the scale value of subjective preference or the period (velocity) of movement of the stimulus had an influence on the EEG. The pairs were selected as pair **1** [period $T = 1.0$ and 0.4 s] and pair

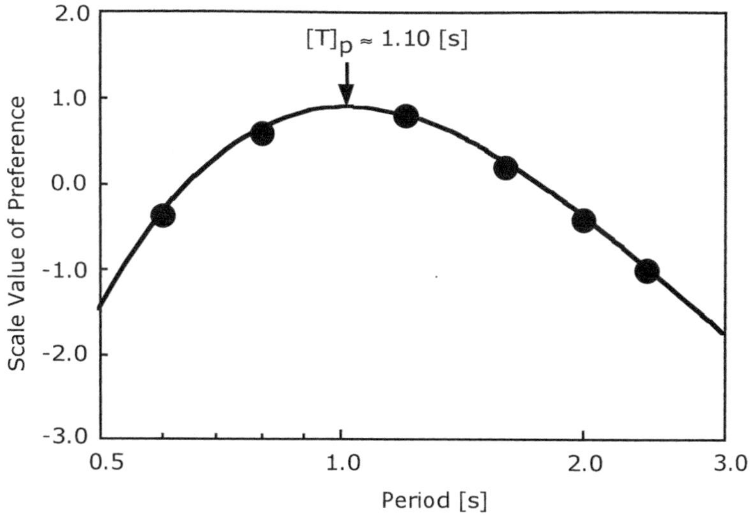

Fig. 3.3 A viewer preference for the rate of visual oscillatory motion. The *arrow* indicates the most preferred period $[T]_p \sim 1.1$ s

2 [period $T = 1.0$ and 4.0 s]. Eight 22- to 26-year-old subjects participated in this study. All had normal or corrected-to-normal vision.

The EEG was recorded using six silver electrodes located at T3, T4, T5, T6, O1, and O2 according to the 10–20 International Electrode Placement System (see Fig. 3.6). The reference electrode was placed at Fz (midline electrodes 30 % of distance from nasion to inion). The ground electrode was placed on the forehead. The recorded data were filtered: 4–8 Hz (theta), 8–13 Hz (alpha), and 13–30 Hz (beta). The integration interval was set at 2.5 s in the CCF analysis as well as in the ACF analysis. An example of a normalized CCF is demonstrated in Fig. 3.1. Results show that the stimulus with the period of 1.0 s induced a longer value of τ_e than did those with periods of 0.4 or 4.0 s.

The values of $|\phi(\tau)|_{max}$ extracted from the CCF were also analyzed to estimate the degree of correlation between cortical responses at different positions. The normalized CCF between the brain waves measured at electrode site O1 (reference electrode) and those measured at the other electrode sites (test electrodes) was analyzed because the difference between the values of τ_e for the two conditions was "the greatest at O1 (left hemisphere)" in the ACF analysis as shown in Fig. 3.4. Significant effects of the stimulus condition and electrode site were reconfirmed in the alpha wave range only when both pairs were presented ($p < 0.01$); namely, no significant effects from the stimulus conditions were detected in the theta and beta ranges. As shown in Fig. 3.5, results reconfirmed that the stimulus with the period of 1.0 s had a greater value of $|\phi(\tau)|_{max}$ in the alpha range than did the stimuli with periods of 0.4 or 4.0 s ($p < 0.01$). As shown in the figure, the values of $|\phi(\tau)|_{max}$ are

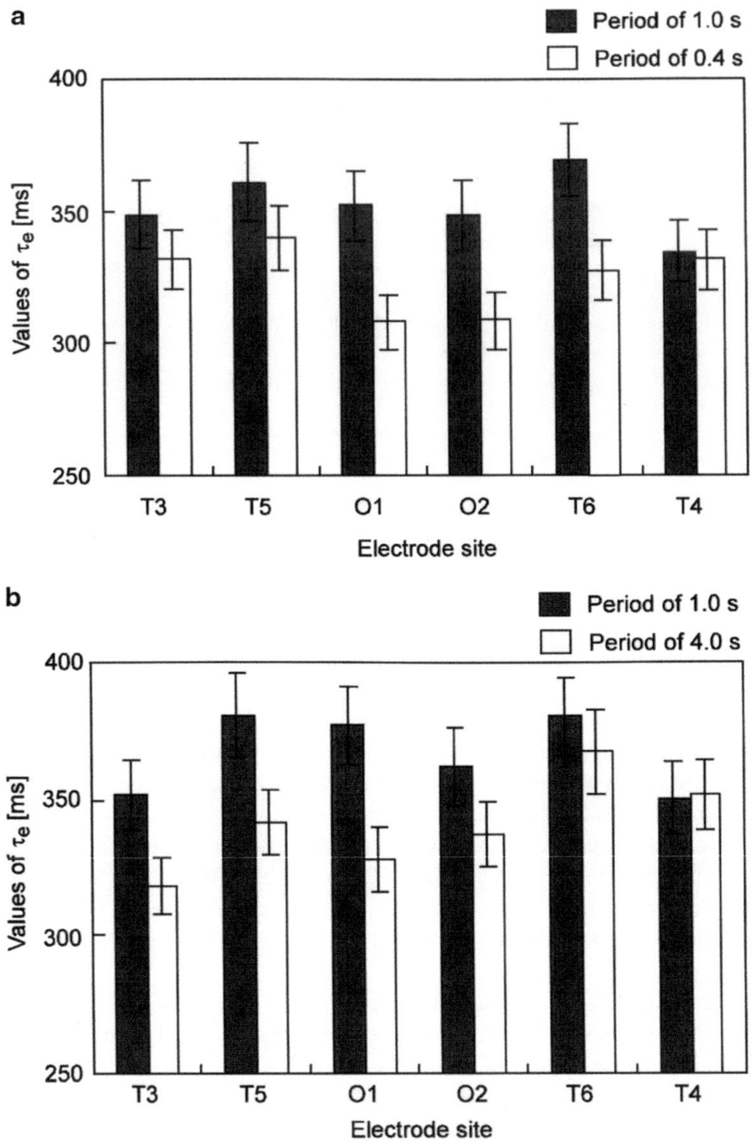

Fig. 3.4 Averaged effective durations τ_e of EEG alpha rhythms range in response to a change of period at different electrode sites. Presentation of (**a**) pair 1 ($T = 1.0$ and 0.4 s) and (**b**) pair 2 ($T = 1.0$ and 4.0 s). The preferred period T is 1.0 s. *Error bars* indicate the standard error of the mean

greater in the posterior temporal and the occipital areas (T5 and O2, respectively) than at the other sites for both pairs 1 and 2.

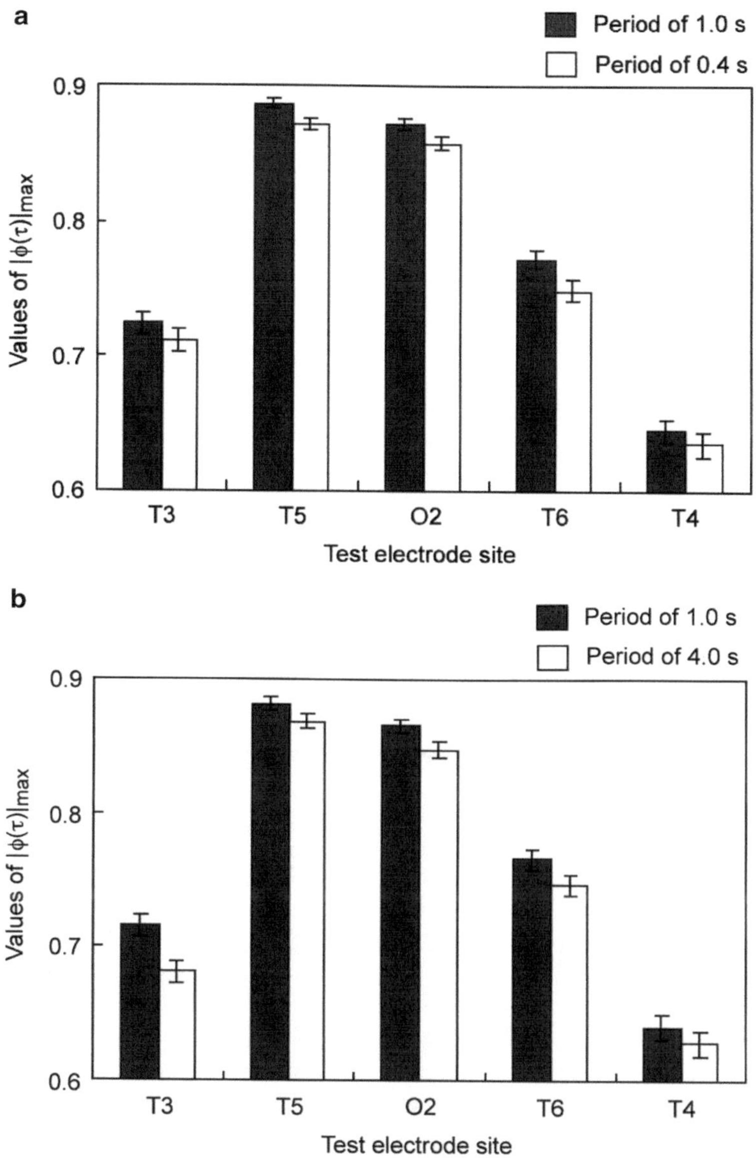

Fig. 3.5 Crosscorrelation analysis of EEG alpha band signals at electrode occipital site O1 and five other sites. Averaged values of maximal crosscorrelation magnitude of $|\phi(\tau)|_{max}$ in response to a change of period under presentation of (**a**) pair 1 ($T = 1.0$ and 0.4 s) and (**b**) pair 2 ($T = 1.0$ and 4.0 s). The preferred period T is 1.0 s. *Error bars* indicate the standard error of the mean

The differences between the scale values of the stimuli with the period of 1.0 s and the stimuli with periods of 0.4 or 4.0 s were not the same for all individual subjects, because the stimuli were selected on the averaged scale value of preference obtained from results of previous subjective preference tests. Results indicate that the value of τ_e in the alpha range was affected by subjective preference, but not by the period of the stimulus itself. The value of τ_e of the ACF in the alpha range, which indicates the degree of persistence of the EEG alpha wave, is prolonged at the preferred condition. This may be caused by the brain repeating the rhythm in the alpha range, which reflects temporal behavior in the cortical area.

This tendency has been found also in previous studies as well as in the effects of varying Δt_1, T_{sub} of the sound field, and in varying the period of the flickering light. In the previous study on the relationship between EEG alpha waves and subjective preference with changes of the period of the flickering light, the value of τ_e was longest in the occipital area. In the current study, however, the value of τ_e was longest not in the occipital area but in the posterior temporal area. This may be a result of experimental conditions: use of a fixed LED in the previous experiment compared to use of visual motion stimuli in this experiment. A study using human positron emission tomography (PET) found that visual system area V5 is located at the occipital-temporal-parietal border (Zeki et al. 1991). The processing of visual motion stimuli in the human visual cortex area V5 has been investigated by using PET, MEG, and EEG. Probst et al. (1993) found that the locations of cortical activation during the presentation of motion stimuli were more lateral than during the presentation of pattern-reversal (non-motion) stimuli.

A significant effect of the stimulus condition on the values of $|\phi(\tau)|_{max}$ was observed in the alpha range. The value of $|\phi(\tau)|_{max}$ averaged in the alpha range for the stimuli with the period of 1.0 s was significantly greater than for stimuli with periods of 0.4 or 4.0 s (Fig. 3.5). We reconfirmed that the values of $|\phi(\tau)|_{max}$ in the alpha range are related to subjective preference, but not by the period of stimuli itself. The fact that the alpha waves have greater $|\phi(\tau)|_{max}$ values shows that the brain repeats its rhythm in the spatial domain over a wider cortical area in the preferred condition than it does in the less preferred condition. This tendency toward greater $|\phi(\tau)|_{max}$ values in the alpha range was also found in the effect of varying the period of the flickering light as discussed above.

In the current study, because O1 (left hemisphere) was chosen as the reference electrode, the value of $|\phi(\tau)|_{max}$ decreased as the distance between O1 and the other electrodes increased. As shown in Fig. 3.6, values of $|\phi(\tau)|_{max}$ were greater in the posterior temporal and occipital areas, T5 and O2, respectively, than in the other areas.

Thus far, we have analyzed the EEG to determine the relationship between the human brain response and subjective preference for horizontal visual motions of movement over varying periods. Results obtained regarding the temporal and the spatial features of brain rhythms can be summarized as:

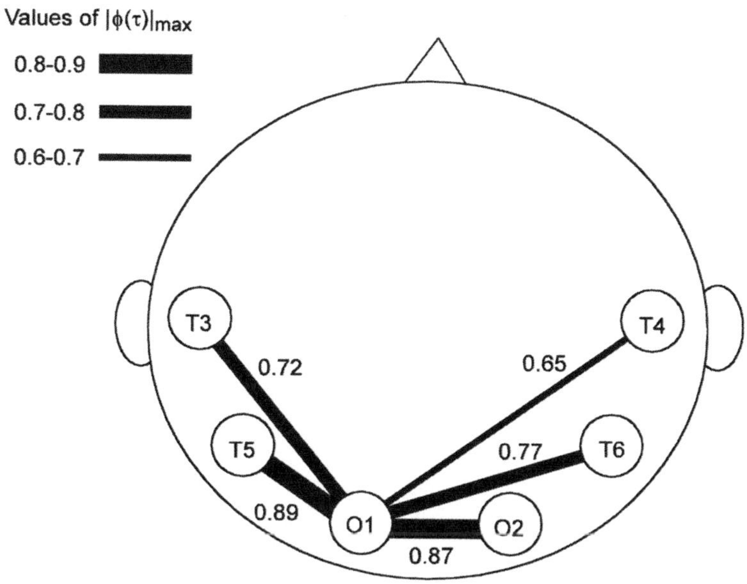

Fig. 3.6 Values of $|\phi(\tau)|_{max}$ of the alpha EEG recording site during the presentation of oscillating moving visual targets. Correlations between signals recorded at different electrodes positions and at O1 are shown for the stimulus pair 1 [$T = 1.0$ and 0.4 s]. The relative thickness of the *bars indicates* the range of the values $|\phi(\tau)|_{max}$

1. The value of τ_e in the alpha wave range for the stimulus with the most preferred period of the movement was significantly longer than for the stimulus with the less preferred periods.
2. This tendency was the greatest at electrode site O1 over the left hemisphere.
3. The value of $|\phi(\tau)|_{max}$ in the alpha range for stimuli with the most preferred period was significantly greater than for stimuli with the less preferred period.

These results reconfirm that the brain repeats the rhythm in the alpha range and that this activity spreads wider over the human brain cortex as a result of the presentation of stimuli with preferred motion rather than with the less preferred motion.

3.2 MEG Correlates of Subjective Preferences for Flickering Light

The MEG was recorded during presentations of the most preferred and less preferred flickering lights alternately during change of the temporal factor. Results showed that (1) the effective duration of the ACF, τ_e, was longer during the preferred condition than during the less preferred conditions and (2) such results were significant in the left hemisphere when the temporal factor was varied.

3.2.1 MEG Correlates of Sinusoidal Flickering Light

To investigate human cortical responses that correspond with subjective preference and hemispheric specialization for visual stimuli, the ACF of the MEG in relation to the period of a flickering light was analyzed (Soeta et al. 2002c). The scale values of individual preference obtained by the PCT together with the averaged scale values of eight subjects are shown in Fig. 3.7. Effects of period on the scale values of preference were examined for all eight subjects by using one-way analysis of variance (ANOVA). Results clearly indicated that the effects of period were significant ($p < 0.01$), and the most preferred period, $[T]_p$, for each subject was estimated by fitting a suitable polynomial curve to a graph on which the scale values were plotted as shown in Fig. 3.8. The preferred period ranged from 0.6 to 2.0 s, and its averaged value was roughly 1.0 s.

Figure 3.9 shows an example of a recorded MEG alpha wave. 16 channels that were located around the occipital area were selected for the ACF analysis and each response to the single stimulus in the pair was analyzed for each subject. The selected 16 channels were divided into three areas as shown in this figure.

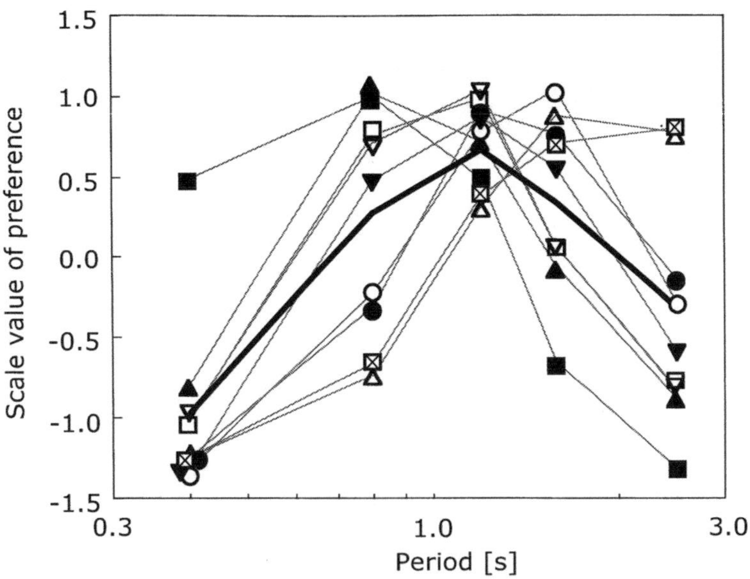

Fig. 3.7 Scale preference values of flickering period T for different subjects. Different symbols represent preferences of different subjects. The *thick black line* shows averaged scale values of preference for all subjects tested

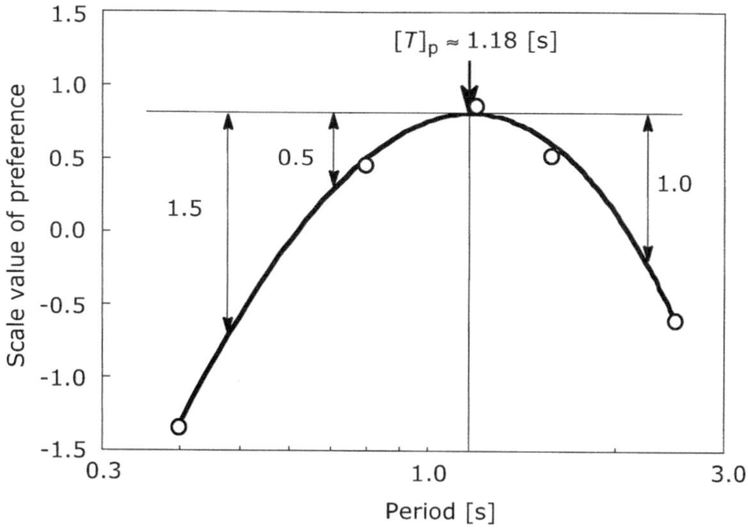

Fig. 3.8 An example of obtaining the most preferred period $[T]_p$ and less preferred periods of the flickering light. $[T]_{0.5} \approx 0.71$, $[T]_{1.0} \approx 2.22$, and $[T]_{1.5} \approx 0.48$ [s] (Suffix number of $[T]$ denotes the differences between scale values of preference)

Significant effects of preference were found on τ_e and ϕ_1 values under all conditions tested. It is clear that the values of τ_e correlated with the value of ϕ_1 ($r = 0.75$, $p < 0.01$). The values of τ_e and ϕ_1 for the most preferred stimuli were larger than for less preferred stimuli for all subjects, as shown in Figs. 3.10 and 3.11, respectively. Such a difference of averaged values of τ_e and ϕ_1 from the left area was significantly larger than of those from the central and right areas ($p < 0.01$). The preferred period of the flickering light clearly induces a much longer τ_e in the alpha wave than the less preferred ones. This tendency for longer τ_e for the preferred period of the flickering light rather than ϕ_1 in the alpha wave was also found in a previous section on EEG (Soeta et al. 2002a). The fact that the brain repeats a similar rhythm under preferred conditions was reconfirmed.

The ratio of high to low preference in terms of the averaged value of τ_e obtained here was small in the range 1.01–1.06, but the difference is significant. This is much smaller than the range derived from the study on EEG, which was 1.18–1.74 (Ando 2009). The EEG results from extracellular volume currents were triggered mainly by the postsynaptic potential. MEG signals are thought to arise from the intracellular currents that flow from the dendritic trees to cell bodies in large numbers (>50,000) of neurons. The MEG and EEG field distributions are mutually orthogonal. Only the current that has a component tangential to the surface of a spherically symmetric conductor produces an exterior magnetic field; the radial source is thus externally silent. Therefore, the radial source might more directly reflect the neural activity patterns associated with preferred visual stimuli. Different results from EEG and MEG for the analysis of human cognition have been discussed previously (Eulitz et al. 1997). Also, numerous studies have reported relationships between

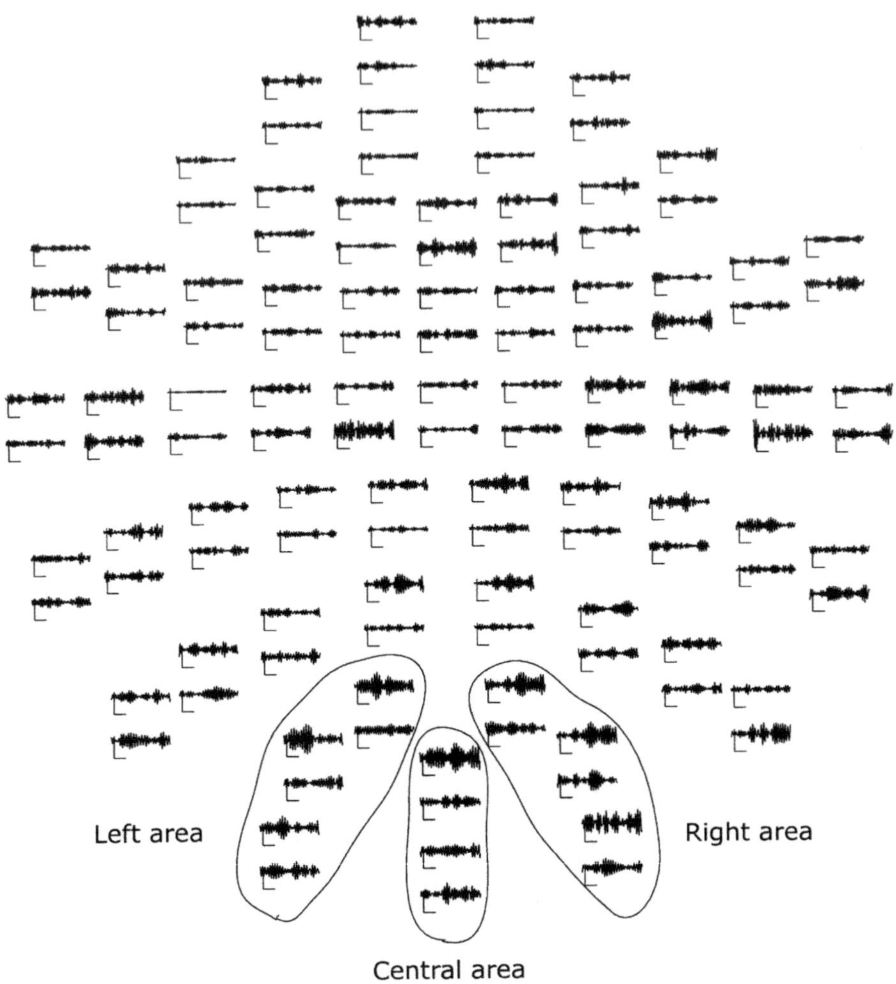

Fig. 3.9 Examples of recorded MEG alpha rhythms. Responses duration was 2.5 s

EEG and MEG coherence and mental processes (Rappelsberger and Petsche 1988; Hinrichs and Machleidt 1992; Petsche 1996). Those studies concentrated on the interchannel relations, for example, synchronization of alpha frequency rhythms.

Here, the values of τ_e and ϕ_1 from the left occipital area were found to be significantly larger than those from the central and right occipital areas. Such a clear tendency was not found in the previous study on EEG discussed above. The significant values τ_e and ϕ_1 in the left hemisphere may reflect the specialization of the human brain, reconfirming specifically the left hemisphere dominance for the temporal factors. It is remarkable that the ratio of high to low preference in terms of the value of τ_e is greater than that of the value of ϕ_1. This indicates that the effective

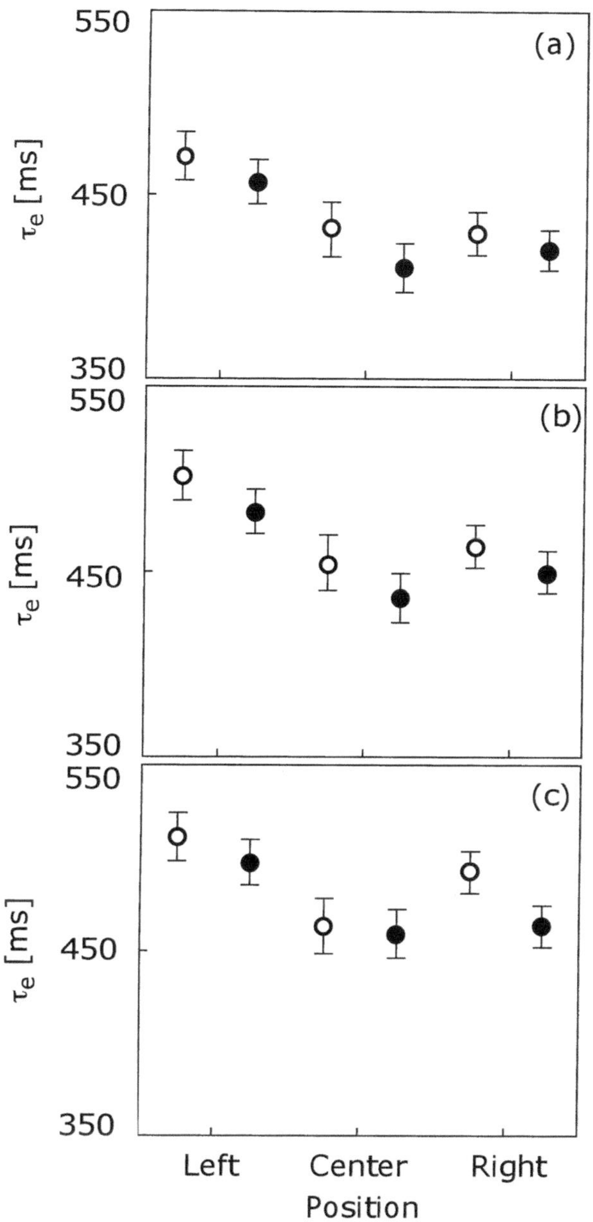

Fig. 3.10 Values of τ_e analyzed from the MEG alpha wave rhythms in response to flickering light recorded from different sensors located in the occipital sensor regions shown in Fig. 3.9. *Error bars* represent 95 % confidence. The difference of the scale value of preference was: (**a**) 1.5. (**b**) 1.0. (**c**) 0.5. ○, higher preference; ●, lower preference

duration τ_e of alpha waves reflects subjective preference better than alpha rhythm regularity ϕ_1.

These MEG experiments investigated human cortical responses corresponding to subjective preferences for flickering lights. The conclusions are:

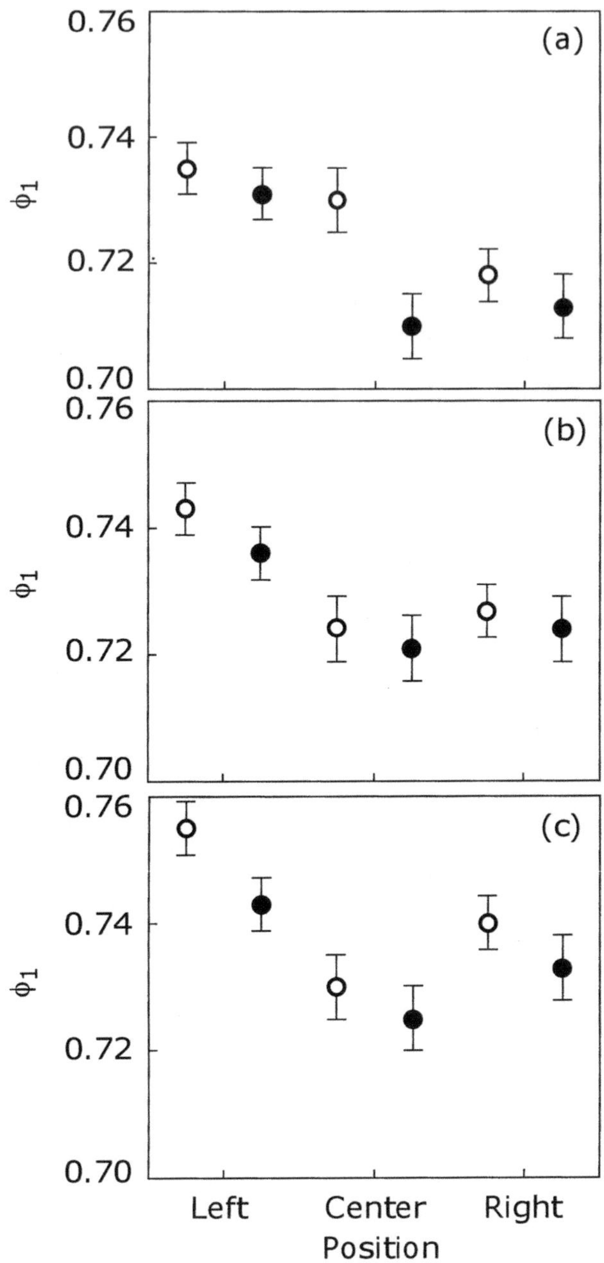

Fig. 3.11 Values of MEG alpha rhythm regularity ϕ_1 in response to flickering light recorded at different occipital sensor regions shown in Fig. 3.9. *Error bars* represent 95 % confidence. The difference of the scale value of preference was: (**a**) 1.5. (**b**) 1.0. (**c**) 0.5. ○, higher preference; ●, lower preference

1. It was reconfirmed that the most preferred flicker periods $[T]_p$ for individuals were 0.6–2 s, with an average value around 1 s.
2. Preferred flicker stimuli evoke significantly longer τ_e of MEG alpha waves than flicker stimuli that are less preferred. There were no clear differences, however, in the average amplitudes of MEG alpha band signals $\Phi(0)$.
3. Averaged values of effective duration τ_e and regularity ϕ_1 of alpha rhythms from the left area were significantly larger than those from the central and right areas.
4. The ratio of high to low preference in terms of effective alpha rhythm duration τ_e is greater than that of the value of alpha rhythm regularity ϕ_1. Thus, the effective duration of alpha rhythms is a better predictor of preference than the regularity of the rhythms.

3.2.2 *MEG Correlates of Subjective Preferences for Fluctuating Flickering Lights*

The most preferred fluctuation of the flickering light was found in the condition of flicker regularity as expressed by $[\phi_1]_p \approx 0.46$. Okamoto et al. (2007) investigated the MEG alpha wave (8–13 Hz) as well as the theta (4–8 Hz) and beta (13–30 Hz) waves over the entire head in response to differing degrees of fluctuation ϕ_1 in the flickering light signal.

Results obtained are as follows:

1. Only in the alpha wave range of the MEG was it found that the values of τ_e of the MEG signals in the alpha range for the most preferred stimuli are longer when compared with those of the relatively less preferred stimuli.
2. This tendency is most pronounced in MEG signals from left occipital sensors (left hemisphere neural responses) where preferred flicker fluctuations that change the regularity of alpha rhythms also increase their effective duration.

3.3 Hemispheric Specialization in Vision

Subjective preference for sound and vision has been described by temporal and spatial factors, which were identified by studies in psychology and neuroscience (Ando 2009). The temporal and spatial factors that can be extracted from the correlation functions existing in the auditory and visual perceptual mechanisms have enabled establishing that the temporal factors of sound and vision are associated with the left cerebral hemisphere and the spatial factors are associated with the right hemisphere. In addition, brain responses related to the scale values of subjective preferences are induced mainly in respective hemispheres. Such a unified science of human perceptual sensations and subjective preferences can be applied for any design practice, in which the goal is to optimize the experience of the design output.

Table 3.1 summarizes our experimental EEG and MEG findings related to hemispheric lateralization of neuronal responses with respect to temporal factors of the visual field (flickering lights and oscillatory horizontal movements). It has been reported that the left cerebral hemisphere is greatly concerned with linear, sequential modes of thinking, such as speech and calculation. On the other hand, it is well known that the right hemisphere tends to perceive space in multiple-dimensional and nontemporal terms for the visual field (Sperry 1974; Davis and Wada 1974; Galin and Ellis 1975; Levy and Trevarthen 1976).

Relationships between alpha rhythms in EEG and MEG recordings and subjective preferences for various aspects of visual stimuli were investigated (Soeta et al. 2002a, b, c; 2003, 2005; Okamoto et al. 2003, 2007). The effective durations of these alpha rhythms were always longest in all subjects ($p < 0.01$) when the preferred flicker rate was presented. Remarkably, averaged values of τ_e and ϕ_1 extracted from the ACF of the MEG alpha wave from the left area were significantly larger than those extracted from the central and right areas.

When the single circular target was moved in the horizontal direction, effects of the subjective preference and measurement position on values of τ_e, $\Phi(0)$, and ϕ_1 extracted from the ACF of the MEG alpha wave (eight subjects) were examined. Effective durations were longer, with larger τ_e, and the alpha rhythms more regular, with larger ϕ_1, for the preferred stimuli compared to the less preferred stimuli ($p < 0.01$). Obviously, the value of τ_e correlated relatively highly with the value of ϕ_1 but only very weakly with the value of $\Phi(0)$. As a conclusion, the averaged values of τ_e and ϕ_1 from the left (O1) area were significantly longer than those from the central and right areas (O2).

Analysis was also performed on the MEG signals from the entire head in response to visual stimuli in which the temporal factor ϕ_1, which reflects the temporal regularity of repeating waveform patterns, was changed using different

Table 3.1 Left hemisphere specialization observed in EEG and MEG alpha waves in relation to the temporal factors of the visual field associated with the left hemisphere

Temporal factors changed	EEG, ACF τ_e value of α-wave	EEG, CCF $\|\phi(\tau)\|_{max}$ of α-wave	MEG, ACF τ_e value of α-wave
Period of the flickering light, *T*	$L > R$[a] (SW)[b]	L & R[c] (SW)	$L > R$ (SW)
Period of horizontal movement of target, T	$L > R$ (SW)	–	–

It is worth noticing that the spatial factors have been reported as associated with the right hemisphere

[a]The ratio of τ_e values of the α-wave in the EEG, from high to low preference, increased significantly in the left hemisphere

[b]Sinusoidal wave used to control the period, T

[c]The $|\phi(\tau)|_{max}$ value of the α-wave in the EEG increased over a wide area in both hemispheres, when the scale value of subjective preference was high. The similar repetitive feature in the α-wave over a wide area of the brain relates to the preferred condition in vision

bandwidths of flicker noise centered on about 1 Hz (Okamoto et al. 2007). The results showed that the τ_e values of the alpha rhythm observed around the left occipital area were significantly larger for the most preferred stimuli than for the less preferred stimuli. This tendency indicates that the stimuli in the preferred condition, with regard to changes of the temporal factor in the visual signal, increase the stability of the alpha rhythm around the left occipital area. The MEG study also revealed that the values of τ_e and ϕ_1 from the left occipital area were significantly larger than those from the central and right occipital areas, but such tendencies were *not found* in the EEG study.

The study on the relationship between the EEG response and the subjective preference by varying the period of the horizontal movement of the single target showed the following: The value of τ_e of the alpha waves for stimulus at the most preferred condition was longer than for the stimulus at the less preferred conditions. This tendency was at maximum at O1 in the left hemisphere. The value of $|\phi(\tau)|_{max}$ of the alpha waves for the stimulus in the most preferred oscillatory movements was greater than for the stimulus in the less preferred conditions. Considering this fact together with the similar repetitive feature in the alpha wave increase, the brain repeats the "alpha rhythm" in the time domain, and this activity spreads wider over the area of the brain cortex at the preferred stimulus condition.

References

Ando Y (2009) Auditory and visual sensations. Springer, New York

Davis AE, Wada JA (1974) Hemispheric asymmetry: frequency analysis of visual and auditory evoked responses to non-verbal stimuli. Electroencephalogr Clin Neurophysiol 37:1–9

Eulitz C, Eulitz H, Elbert T (1997) Differential outcomes from magneto- and electroencephalography for the analysis of human cognition. Neurosci Lett 227:185–188

Galin D, Ellis RR (1975) Asymmetry in evoked potentials as an index of lateralized cognitive processes: relation to EEG alpha asymmetry. Neuropsychologia 13:45–50

Hinrichs H, Machleidt W (1992) Basic emotions reflected in EEG-coherences. Int J Psychophysiol 13:225–232

Levy J, Trevarthen C (1976) Metacontrol of hemispheric function in human split-brain patient. J Exp Psychol Hum Percept Perform 2:299–312

Okamoto Y, Soeta Y, Ando Y (2003) Analysis of EEG relating to subjective preference of visual motion stimuli. J Temporal Des Architect Environ 3:36–42. http://www.jtdweb.org/journal/

Okamoto Y, Nakagawa S, Yano T, Ando Y (2007) An MEG study of cortical responses in relation to subjective preference for different regularities of a fluctuating light. J Temporal Des Architect Environ 7:10–18. http://www.jtdweb.org/journal/

Petsche H (1996) Approaches to verbal, visual and musical creativity by EEG coherence analysis. Int J Psychophysiol 24:145–159

Probst T, Plendl H, Paulus W, Wist ER, Scherg M (1993) Identification of the visual motion area (area V5) in the human brain by dipole source analysis. Exp Brain Res 93:345–351

Rappelsberger P, Petsche H (1988) Probability mapping: power and coherence analysis of cognitive processes. Brain Topogr 1:46–54

Soeta Y, Uetani S, Ando Y (2002a) Relationship between subjective preference and alpha wave activity in relation to temporal frequency and mean luminance of a flickering light. J Opt Soc Am A 19:289–294

Soeta Y, Uetani S, Ando Y (2002b) Propagation of repetitive alpha waves over the scalp in relation to subjective preferences for a flickering light. Int J Psychophysiol 46:41–52

Soeta Y, Okamoto Y, Nakagawa S, Tonoike M, Ando Y (2002c) Autocorrelation analyses of MEG alpha waves in relation to subjective preference of a flickering light. NeuroReport 13:527–533

Soeta Y, Ohtori K, Ando Y (2003) Subjective preference for movements of a visual circular stimulus: a case of sinusoidal movement in vertical and horizontal directions. J Temporal Des Architect Environ 3:70–76. http://www.jtdweb.org/journal/

Soeta Y, Mizuma K, Okamoto Y, Ando Y (2005) Effects of the degree of fluctuation on subjective preference for a 1 Hz flickering light. Perception 34:587–593

Sperry RW (1974) Lateral specialization in the surgically separated hemispheres. In: Schmitt FO, Worden FC (eds) The neurosciences: third study program. MIT Press, Cambridge, Chapter 1

Zeki S, Watson JDG, Lueck CJ, Friston KJ, Kennard C, Frackowiak RSJ (1991) A direct demonstration of functional specialization in human visual cortex. J Neurosci 11:641–649

Chapter 4
Temporal and Spatial Percepts of Sound and Vision

4.1 Definitions of Temporal and Spatial Objective Factors

4.1.1 *Temporal Factors for Sound and Vision*

The mechanism that is deeply embedded in auditory and visual systems is the autocorrelation function (ACF), which is defined by

$$\Phi_p(\tau) = \lim_{T \to \infty} \frac{1}{2T} \int_{-T}^{+T} p'(t)p'(t+\tau)\mathrm{d}t \tag{4.1}$$

where $p'(t) = p(t)*s(t)$, $s(t)$ is the sensitivity of the ear. For the auditory system, $s(t)$ is mainly formed by the physical system between the ear entrance and the oval window (Ando 1985, 1998). For the sake of convenience, $s(t)$ can be chosen as the impulse response of an A-weighted network. And $p(t)$ is a one-dimensional sound and visual signal in the time domain. The normalized ACF is defined by

$$\phi p(\tau) = \Phi p(\tau)/\Phi_p(0) \tag{4.2}$$

There are four significant objective factors, which can be extracted from the ACF of a signal:

1. Energy represented at the origin of the delay, $\Phi_p(0)$.
2. As shown in Fig. 4.1, the width of amplitude $\phi(\tau)$, around the origin of the delay time ($\tau = 0$) defined at a value of 0.5, is $W_{\phi(0)}$. It is noteworthy that $\phi(\tau)$ is an even function.
3. Fine structure, including peaks and delays. For instance, τ_1 and ϕ_1 are the delay time and the amplitude of the first peak of the ACF, τ_n and ϕ_n being the delay time and the amplitude of the nth peak. Usually, there are certain correlations between τ_1 and τ_{n+1} and between ϕ_1 and ϕ_{n+1}, so that significant factors are τ_1 and ϕ_1.

Y. Ando, *Brain-Grounded Theory of Temporal and Spatial Design*,
DOI 10.1007/978-4-431-55891-0_4

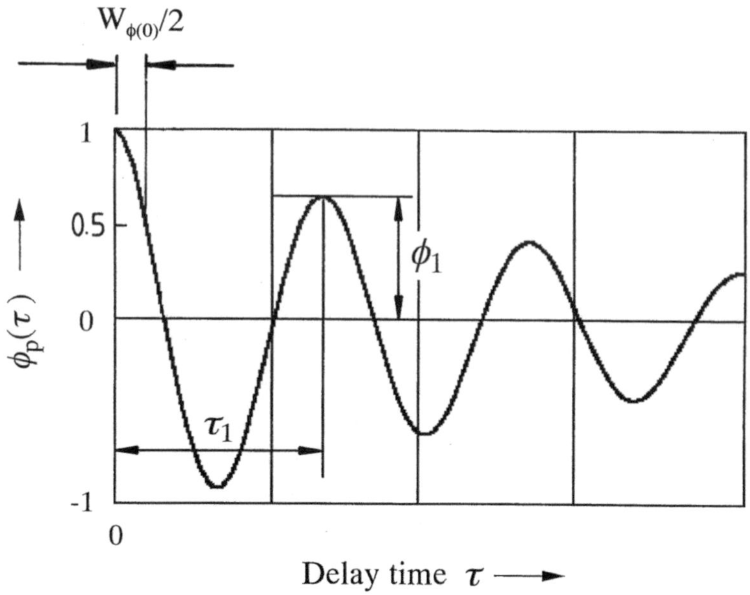

Fig. 4.1 Definition of three temporal factors, $W_{\phi(0)}$, τ_1, and ϕ_1, extracted from the running autocorrelation function (ACF) of sound signals

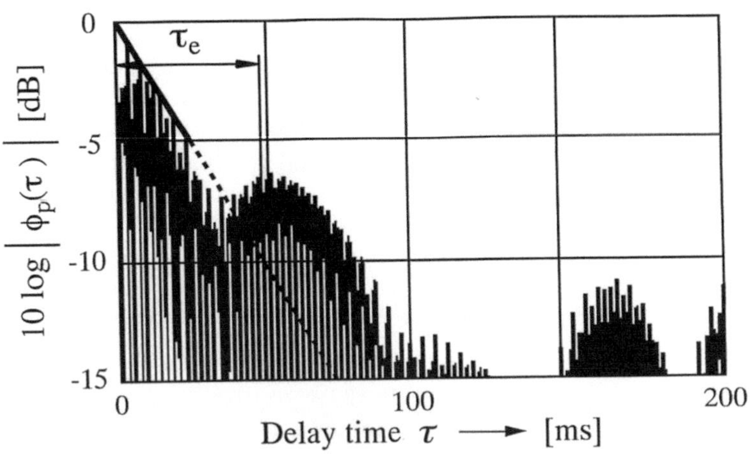

Fig. 4.2 Method of obtaining the effective duration of the running ACF, τ_e. The ten-percentile (−10 dB) delay of its envelope

4. The effective duration of the envelope of the normalized ACF, τ_e, which is defined by the ten-percentile delay and which represents a repetitive feature or "reverberation" containing the source itself (Fig. 4.2).

For example, the ACF of any sinusoid (pure tone) having any phase is a zero-phase cosine function of the same frequency. Since its waveform and ACF envelope are flat, with a slope of zero, its effective duration τ_e is infinite. The ACF of white noise with infinite bandwidth is the Dirac delta function $\delta(\tau)$, which has an infinite slope. This means that the signal has an effective duration that approaches zero (no temporal coherence). As the bandwidth of the noise decreases, the effective duration (signal coherence) increases.

4.1.2 Spatial Factors Extracted from the Interaural Crosscorrelation Function (IACF) of the Sound Field

The IACF for the sound pressures for the temporal window $2T$, $p(t)_{l,r}$, arriving at each of the two ear entrances is given by

$$\Phi lr(\tau) = \frac{1}{2T}\int_{-T}^{+T} f'_l(t) f'_r(t+\tau)\,dt \tag{4.3}$$

where $f'_l(t)$ and $f'_r(t)$ are obtained by signals $f'_{l,r}(t)$ after passing through the A-weighted network, which corresponds to the ear's sensitivity $s(t)$. The normalized IACF in the possible range of maximum interaural delay times is given by

$$\phi_{lr}(\tau) = \Phi_{lr}(\tau)/[\Phi_{ll}(0)\Phi_{rr}(0)]^{1/2}, \quad -1\text{ ms} < \tau < +1\text{ms} \tag{4.4}$$

where $\Phi_{ll}(0)$ and $\Phi_{rr}(0)$ are the ACFs at $\tau = 0$ or sound energies arriving at the left- and right-ear entrances, respectively, and τ is the interaural time delay possible within plus and minus 1 ms.

From the IACF analysis, four spatial factors are determined (Fig. 4.3):

1. The magnitude of the interaural crosscorrelation is defined by

$$\text{IACC} = |\phi_{lr}(\tau)|_{max} \tag{4.5}$$

for the possible maximum interaural time delay, i.e., -1 ms $< \tau <$ $+1$ ms for humans.

2. The interaural delay time, at which the IACC is defined is τ_{IACC}.
3. The width of the IACF, defined by the interval of delay time at a value of $\delta = 0.1$ below the IACC, corresponding to the just-noticeable difference (JND) of the IACC, is given by W_{IACC}. Thus, the apparent source width (ASW) may be perceived as a directional range relating to the W_{IACC}. This value mainly depends on the frequency component of the source signals. A well-defined horizontal localization, or directional impression, corresponding to the interaural time delay τ_{IACC} is perceived when listening to a sound with a sharp peak in the

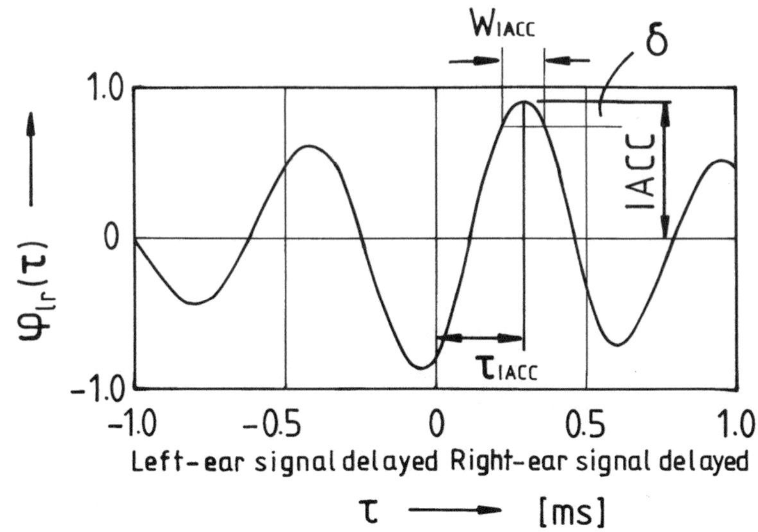

Fig. 4.3 Definition of three spatial factors, IACC, W_{IACC}, and τ_{IACC}, extracted from the interaural crosscorrelation function IACF of sound signals arriving at the two ears

IACF with a short value of $\mathrm{W_{IACC}}$. On the other hand, when listening to a sound field with a low value for the IACC < 0.15, then a subjectively diffuse sound field is perceived (Damaske and Ando 1972).

4. The denominator of Eq. (4.4), which is defined by the LL, is the geometric mean of the sound energies arriving at the two ears.

The spatial sensations of the sound field are described by means of these factors extracted from the IACF. These spatial sensations may be judged immediately when we come into a sound field, because our binaural system may process the IACF in a short temporal window as discussed below. The human process of analyzing the spatial characteristics of a sound field is quite different from the adaptive temporal window for the sound signals, which depend on the effective duration of the ACF of the sound-source signal.

4.1.3 Spatial Factors Extracted from the Spatial Autocorrelation Function (ACF) of the Visual Field

In order to describe a two-dimensional texture, the spatial factors need to be extracted from the ACF of the gray level excluding any color information. Here, we are not discussing binocular signals, which are related to depth perception. The underlying problem is how the human visual system characterizes salient spatial sensations of the texture.

The ACF of a digitalized two-dimensional texture pattern is expressed by

$$\Phi(\Delta x, \Delta y) = \sum_{x=0}^{M-1}\sum_{y=0}^{N-1} p(x, y)p(x + \Delta x, y + \Delta y)/MN \tag{4.6}$$

where $p(x, y)$ is the digitalized input signal and $p(x + \Delta x, y + \Delta y)$ is the spatially shifted version of the input. Integers "M" and "N" refer to the data size in horizontal and vertical directions. In this study, these values were selected as $M = N = 256$. The normalized ACF is given by

$$\phi(\Delta x, \Delta y) = \Phi(\Delta x, \Delta y)/\left(\sum_{x=0}^{M-1}\sum_{y=0}^{N-1} p(x, y)^2/MN\right) \tag{4.7}$$

The denominator in Eq. (4.7) is the maximum value of the ACF representing the power, or energy, of one of the factors of contrast in the image. Thus, the normalized ACF has the maximum value of 1.0 at the origin, $\Delta x = \Delta y = 0$.

The ACF contains the same information as the power density spectrum. We use the ACF here because it allows us to most simply and directly obtain the significant primary factors involved in spatial texture perception in a manner similar to auditory perception (Ando 2008, 2009). Remarkably, the spatial ACF of the pattern has a dominant periodicity that corresponds to the reciprocal of its fundamental frequency. Even when the fundamental frequency component is subtracted from the stripe pattern, due to the missing fundamental, we can still see the fundamental periodicity (Henning et al. 1975). The frequency domain analysis fails to explain this percept, because the fundamental frequency component is absent. Instead, the ACF of this missing fundamental pattern maintains a dominant periodicity at the fundamental. The perceptual mechanism of the pattern is still under discussion (Badcock and Derrington 1989; Hammett and Smith 1994); however, it indicates that the ACF is effective for detecting the spatial periodicity of patterns.

As shown in Fig. 4.4, the ACF decays from the origin outward. To simplify the calculation, only two one-dimensional ACFs along the x- and y-axes from the origin were considered. This is because the most natural texture is considered to be *isotropic*, in which all orientations occur with the same probability. Consequently, the ACF could be assumed to be circular symmetric. Even for the *anisotropic* texture, at least one of four directional ACFs can hold the periodic structure. The four factors determined are

1. $\Phi(0,0)$: the autocorrelation at the origin, $\Delta x = \Delta y = 0$
2. δ_1: the displacement of the first maximum peak in the ACF
3. ϕ_1: the amplitude of the first maximum peak
4. δ_e: the effective range or effective spatial duration of the ACF (defined as the displacement at which the envelope of the normalized ACF decayed at 0.1)

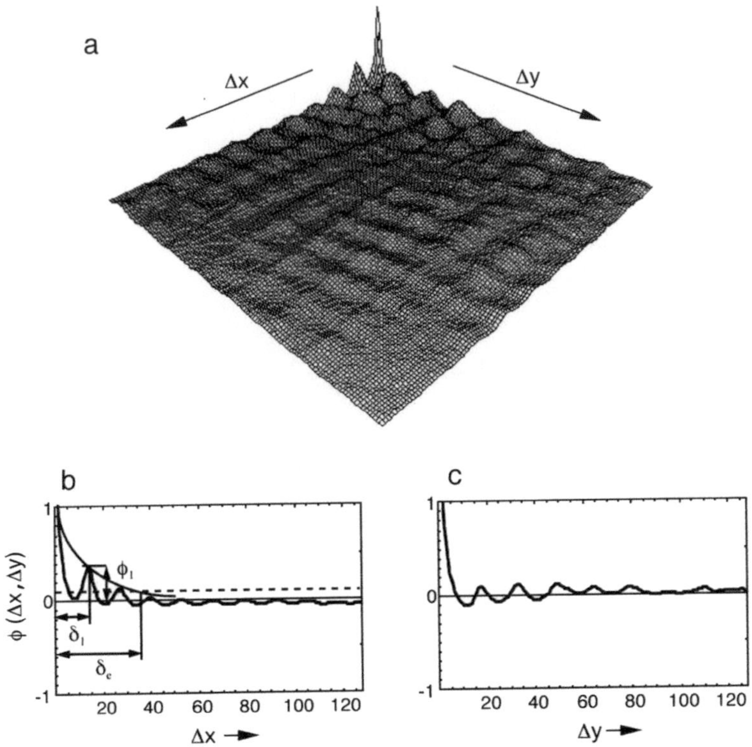

Fig. 4.4 An example of the spatial ACF analyzed and factors extracted from the ACF. (**a**) Two-dimensional ACF of texture D3 in Fig. 4.17a.. (**b**) One-dimensional spatial ACF for the x direction. (**c**) One-dimensional spatial ACF for the y direction. Symbol $\phi(\Delta x,\Delta y)$ signifies the spatial ACF for the spatial intervals Δx and Δy, ϕ_1 is the amplitude of the shortest major spatial interval, δ_1 is the length of shortest major spatial interval, and δ_e is the effective range of significant intervals. By comparing the amplitudes and shortest spatial intervals in both directions, we extracted three factors from the ACF for the x direction only. Note that the x direction indicates more complicated spatial structures throughout the texture applied here

The factor $\Phi(0,0)$ corresponding to the energy of the image was calculated from the two-dimensional ACF, i.e., the denominator in Eq. (4.7), and it is shown in the dB scale $[10\log_{10}\Phi(0)]$. The algorithm calculating the other three factors is as follows. First, the directional ACFs along the x- and y-axes are extracted from the two-dimensional ACF. Then, the first maximum peak value in both directions is figured out. From the selected one-dimensional ACF, the displacement and the amplitude of the maximum peak are picked out as δ_1 and ϕ_1 (Fig. 4.4). The value of effective spatial duration δ_e is determined from the cross point of the envelope at 0.1, that is, by the point at which the normalized spatial ACF decays to 10 % of its zero-lag maximal value, as shown in the figure. This is done by exponential fit to normalized ACF peaks above 0.1. When the given ACF has no peaks above 0.1, only the initial decay rate was fitted.

The displacement and the amplitude of the maximum peak (δ_1 and ϕ_1) are related to the periodicity of the image. The value of δ_1 is a reciprocal of the fundamental frequency, and ϕ_1 is the strength of the harmonic components. The value $1/\delta_1$ represents the spatial pitch, and the value of ϕ_1 represents the strength of the spatial pitch (Fig. 4.4). The image width (cm) and the distance (cm) calculate the visual angle in degrees.

4.1.4 Temporal Windows of Signals for Autocorrelation Processing

(a) In analyzing the running temporal ACF of a signal, we need to know the temporal integration time, or the so-called temporal window 2T of the signal.

Successive loudness judgments in pursuit of the running listening level LL have been conducted (Mouri et al. 2001). Results are shown in Fig. 4.5. The recommended temporal window, or signal duration $(2T)_r$, to be analyzed is approximately given by

$$(2T)_{\mathrm{r}} \approx 30\,(\tau)_{\mathrm{min}} \tag{4.8}$$

where $(\tau_e)_{min}$ is the minimum value of effective durations obtained by analyzing the running ACF. The temporal integration window should thus be about 30 times the minimum effective duration of the signal.

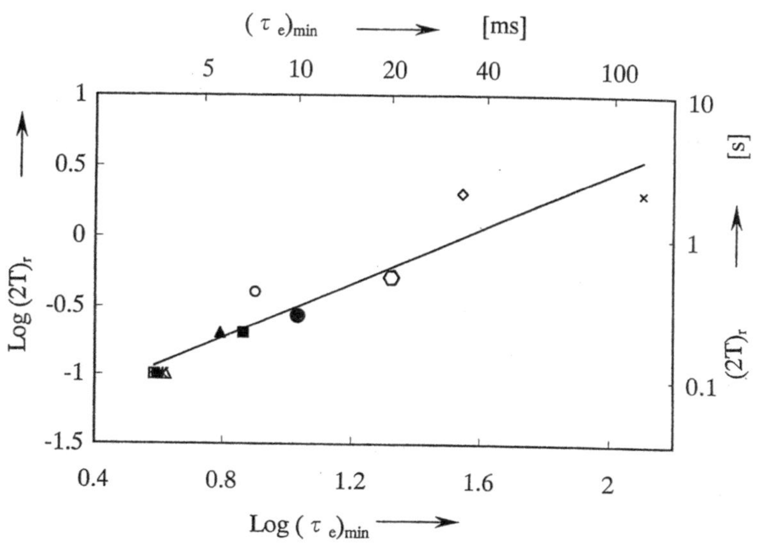

Fig. 4.5 Recommended signal duration $(2T)_r$ of signals to be analyzed by the ACF (Mouri et al. 2001)

(b) For analyzing the running IACF, the temporal window $2T$ should be selected according to the location of the sound source.

When the source of white noise is moving around a listener, then the recommended window interval $[2T]_{RI}$ for analyzing the running IACF should be short, as given by (Ando 2009):

$$[2T]_{RI} \approx 30 \text{ ms} \tag{4.9a}$$

For a signal source that is fixed in location, the window of the IACF should be

$$[2T]_{RI} \approx 1.0 \text{ s} \tag{4.9b}$$

4.2 Temporal and Spatial Percepts of Sound

Based on the high-level signal processing model of the central auditory system shown in Fig. 2.11, primary temporal and spatial percepts can be described by the temporal and spatial factors, respectively. Temporal percepts of sound are pitch, monaural loudness, timbre, and duration, which are formulated by the temporal factors extracted from the autocorrelation function (ACF) (Table 4.1). In contrast, spatial percepts of the sound field, that is, localization of sound, apparent source width (ASW), and subjective diffuseness, which is deeply associated with envelopment, are formulated by the spatial factors extracted from the interaural crosscorrelation function (IACF) of sound signals arriving at the two ear entrances (Table 4.2). Figure 4.6 summarizes the auditory temporal and spatial percepts, which may be formulated by the temporal factors extracted from the ACF of the sound signal and the spatial factors extracted from the IACF.

Table 4.1 Auditory temporal sensations in relation to the temporal factor extracted from the ACF

Temporal factors		Temporal sensations[a]			
		Loudness	Pitch	Timbre	Duration sensation
ACF	$\Phi_p(0)$	(X)			–
	τ_1	X[b]	X ($F = 1/\tau_1$)		X
	ϕ_1	X[c]	X (strength of pitch)	–	x
	τ_e	X			
	$W_{\phi(0)}$			X	
	D				X

[a]It is noteworthy that temporal sensations are associated with the left hemisphere, so that temporal sensations may be additionally affected by two temporal factors of the sound field: Δt_1 and T_{sub}. For example, Sayles and Winter (2008) have shown that pitch is influenced by the reverberant energy ratio equivalent to the value A that depends on the distance between the sound source and the receiving position

[b]X: Major factors describing corresponding temporal sensations

[c]X: Attention to the fact that factors ϕ_1 and τ_e are mutually related

Table 4.2 Auditory spatial sensations in relation to the factor extracted from the IACF

Spatial factors		Spatial sensations		
		ASW	Subjective diffuseness	Localization in the horizontal plane
IACF	LL[a]	X[b]	X	–[c]
	τ_{IACC}	–	–	X
	W_{IACC}	$(W_{IACC})^{1/2}$	–	–
	IACC	$(IACC)^{3/2}$	$(IACC)^{3/2}$	X

Note that factors for localization in the median plane may be extracted from the ACF (Sato et al. 2001)

[a]$LL = 10 \log [\Phi(0)/\Phi(0)_{reference}]$, where $\Phi(0) = [\Phi_{ll}(0)\ \Phi_{rr}(0)]^{1/2}$

[b]X: Major factors describing corresponding spatial sensations. (The related section indicated in the bracket.)

[c]–: Factors to be examined for their influence/insignificance on the respective sensation

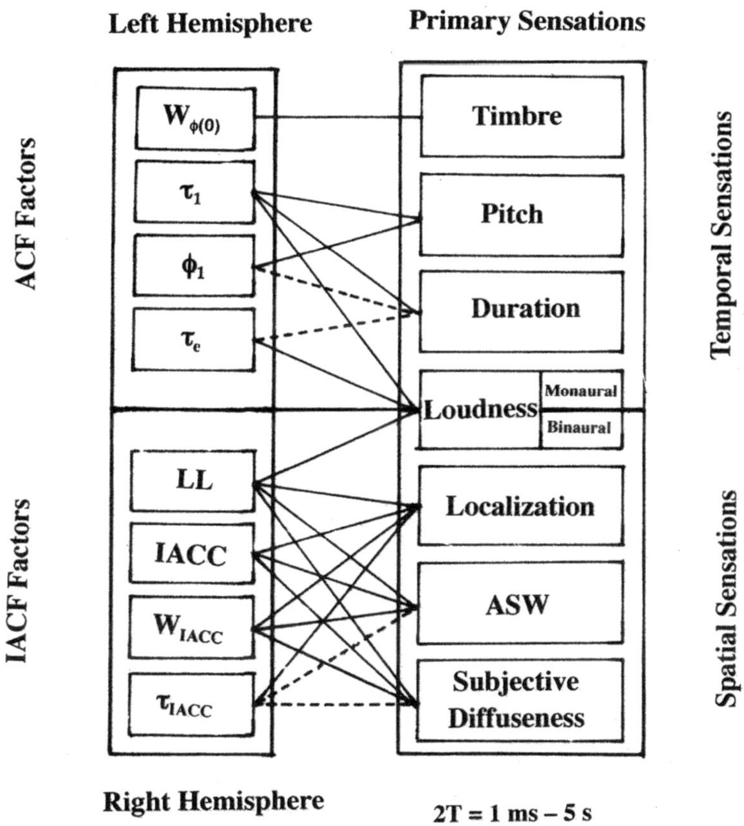

Fig. 4.6 Primary temporal and spatial sensations (percepts), respectively, associated with the left and the right cerebral hemisphere of the sound signal and the sound field

4.2.1 Temporal Sensations of Sound

1. Pitch

The pitch (missing fundamental) of complex tones as a temporal percept associated with the left hemisphere (L) may be expressed by

$$S = S_L = f_L(\tau_1) \approx 1/\tau_1 [\text{Hz}] \tag{4.10a}$$

where τ_1 is extracted from the ACF as defined by previous section ($\phi_1 = 1$).

Pitch-matching tests, comparing pitches of pure and complex tones, have been conducted; the test signals were all complex tones consisting of harmonics 3–7 of a 200 Hz fundamental (600, 800, 1000, 1200, 1600 Hz) (Sumioka and Ando 1996; Ando 2002). Results of the pitch-matching test with five subjects in the conditions of in-phase and random-phase components are shown in Fig. 4.7. The ACFs of both cases are identical and the delay time at the first maximum peak of the NACF, τ_1, equals 5 ms (200 Hz), corresponding to the fundamental frequency. There are no major differences between results of the in-phase and random-phase conditions pertaining to Eq. (4.10a). Pitch matches for all subjects are shown in Fig. 4.8 (Inoue et al. 2001), which illustrates the limitations of the above equation. Whenever the missing fundamental frequency of the stimulus was 500, 1000, or 1200 Hz, more than 90 % of the responses obtained from all subjects under both conditions clustered around the fundamental frequency. When missing fundamentals were 1600, 2000, or 3000 Hz, however, the probability that the subjects matched the frequency of the pure tone to the fundamental frequency was much lower. These results imply that the ACF model is applicable when stimuli have missing fundamentals of 1200 Hz or less as shown in Fig. 4.8 (Inoue et al. 2001). The highest frequency components of the sound signal should be under 5 kHz.

Pitch strength is given by (Ando 2009)

$$S_p = k\Phi_1 \tag{4.10b}$$

where k is the coefficient. The maximum strength is observed at $\phi_1 = 1.0$, and no pitch is perceived when $\phi_1 < \varepsilon$, ε being a small positive number.

2. Loudness

If the sound-pressure level SPL [dB] is measured by a single microphone, then the scale value of "monaural loudness" S associated with the left hemisphere is expressed by

$$S = S_L = f_L(\text{SPL}, \tau_1, \Phi_1, \tau_e, D) \tag{4.11}$$

where τ_1, ϕ_1, and τ_e are factors extracted from the ACF and D is the duration of the sound signal. In a binaural listening condition, SPL being the sound-pressure level

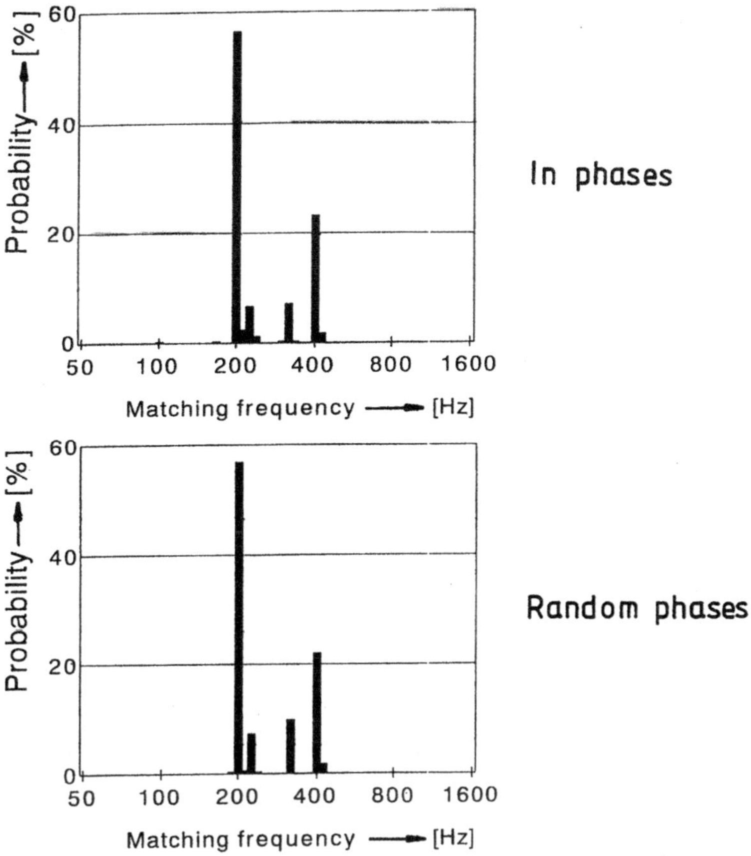

Fig. 4.7 Results of pitch-matching tests for two complex tones of pure-tone components of 600, 800, 1000, 1200, and 1600 Hz in conditions of in phase and random phases

is replaced by the binaural listening level (LL) associated with the right hemisphere, which is obtained by

$$LL = 10\log[\Phi_{ll}(0)\Phi_{rr}(0)]^{1/2}/\Phi_{ref}(0) \tag{4.12}$$

where $\Phi_{ll}(0)$ and $\Phi_{rr}(0)$ are the energies arriving at the left and right ear entrances, respectively. $\Phi_{ref}(0)$ is the reference level of the SPL.

Loudness judgments were conducted with five subjects applying the paired-comparison tests (PCT) by changing the bandwidth, using a sharp filter with a cutoff slope of 2068 dB/octave. Results for the center frequencies of 250, 500, and 1000 Hz are shown in Fig. 4.9a–c. For example, loudness of the band-pass noise produced by such a sharp filter with identical SPL was not constant within the critical band. This is quite a different result from the previous studies (Zwicker et al. 1957). Also, loudness of the pure tone was significantly larger than that of

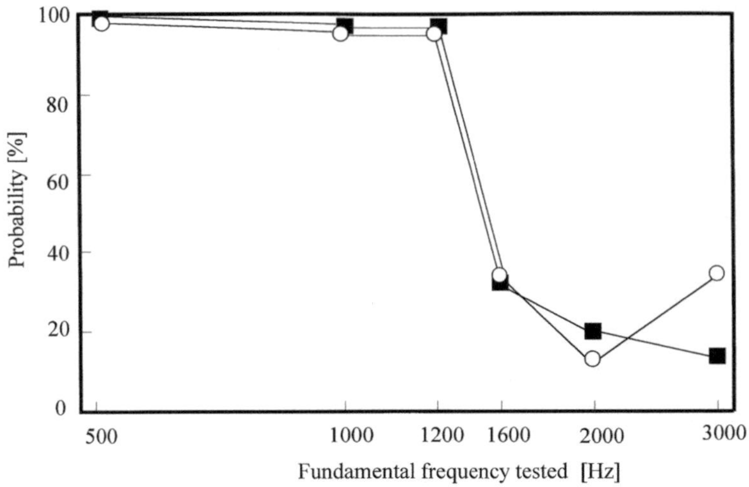

Fig. 4.8 Probability of subjects adjusting a pure tone near the fundamental frequency of complex tones. *Empty circles* are results for two harmonics, and *full squares* are results for three harmonics

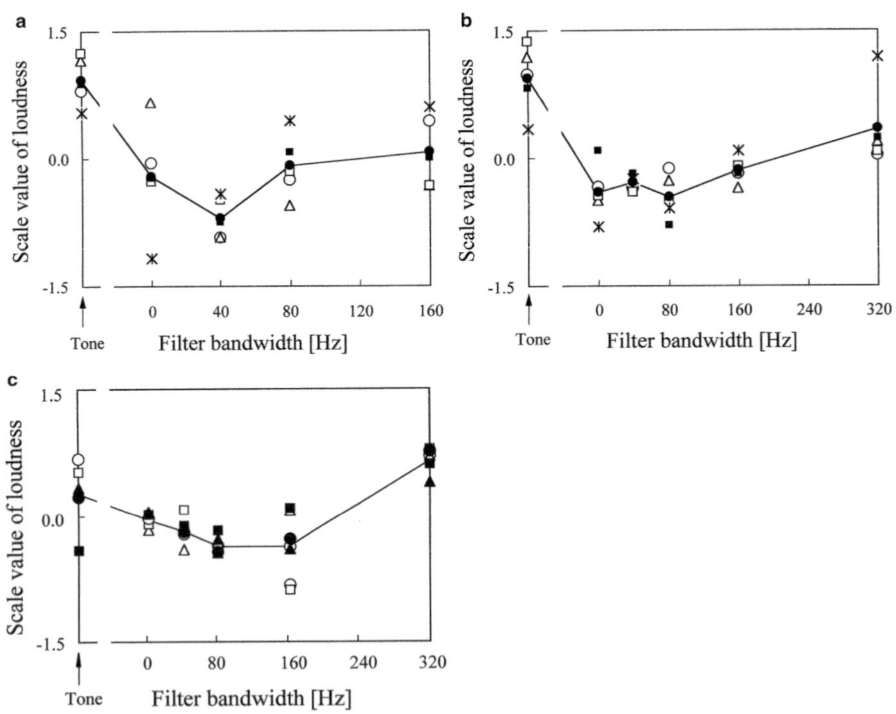

Fig. 4.9 Scale values of loudness as a function of the bandwidth of noise. Different symbols indicate the scale values obtained with different subjects. (**a**) $f_c = 250$ Hz. (**b**) $f_c = 500$ Hz. (**c**) $f_c = 1000$ Hz

filtered noises, and loudness increased with increasing τ_e within the critical band (Sato et al. 2002); therefore,

$$S = S_L = f_L(\tau_1) + f_L(\tau_e) \tag{4.13}$$

This means that if environmental noise contains a pure-tone component or long values of τ_e, such as is perceived in the noise of landing aircraft, then loudness increases and annoyance increases accordingly (Soeta et al. 2004). This is a typical example of why only SPL measurement by the sound level meter does not qualify as being a well-described subjective attribute. Just adding +10 dB to such noise with a coherent component, which is the recommended method, is far too simple and insufficient to produce serious results, because both loudness and annoyance increase as the value of τ_e increases.

3. Duration

The sensation of temporal duration for pure and complex tones has been discussed (Saifuddin et al. 2002). Obviously, the duration percept depends most directly on the physical signal duration, D. In terms of internal auditory representations, we show here that perceived duration also covaries with tone frequency and with its corresponding temporal factor in the ACF, τ_1, which corresponds to pitch, so that

$$S = S_L = f_L(\tau_1, D) = f_L(\tau_1) + f_L(D) \tag{4.14}$$

As Fig. 4.10 shows, apparent stimulus duration (DS) depends primarily on the duration of the signal and secondarily on signal periodicity τ_1 (pure-tone frequency

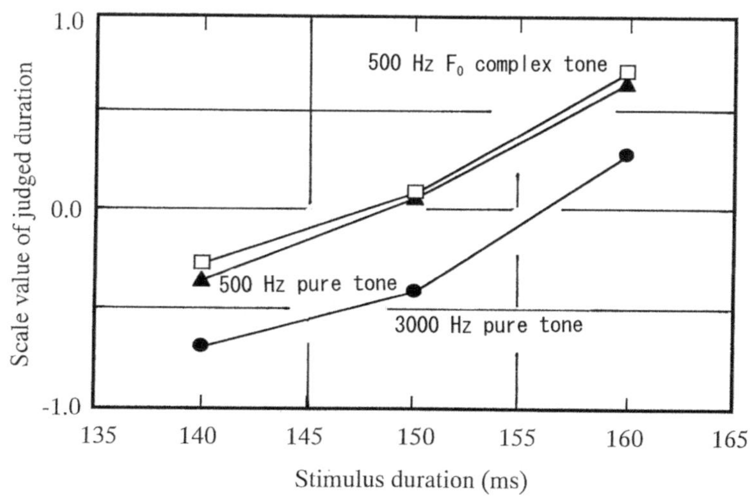

Fig. 4.10 Scale values of duration sensation (DS) by the PCT.□: Complex tone (F0 = 500 Hz) with 3000 Hz and 3500 Hz pure-tone components. ▲: 500 Hz pure tone. ●: 3000 Hz pure tone

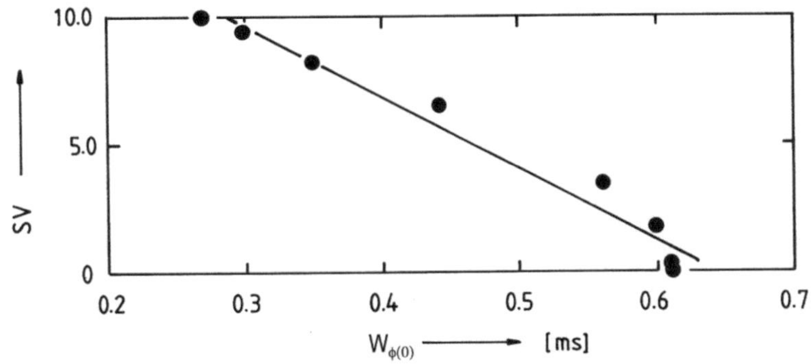

Fig. 4.11 Results of regression analysis for SV and the mean value of $W_{\phi(0)}$

or complex-tone fundamental frequency). Effects of the τ_1 extracted from the ACF on DS are almost the same on the scale value for the pure-tone ($\tau_1 = 2$ ms) and complex-tone ($\tau_1 = 2$ ms) stimuli. The apparent duration DS of the pure-tone stimulus ($\tau_1 = 0.33$ ms = 1/3000 Hz) with the higher pitch is significantly shorter than that of the pure-tone and complex-tone stimuli with the lower pitch ($\tau_1 = 2$ ms = 1/500 Hz). Namely, when the sound signal is lower pitch, then the duration sensation may be longer. Therefore, apparent duration DS can be readily expressed as a function of D and τ_1 for both pure and complex tones.

4. Timbre

Timbre is defined as an aspect of sound quality that is independent of loudness, pitch, and duration, i.e., it is the perceived quality of sound (often described in terms of sound texture or coloration) that distinguishes two notes of equal pitch, loudness, and duration that are played on different musical instruments. As shown in Fig. 4.11, the factor $W_{\phi(0)}$ extracted from the ACF reflects the relative width of the ACF peak at its zero-lag origin. $W_{\phi(0)}$ is defined by the first delay time $\phi(\tau)$ at which the normalized ACF declines to half of its maximal value (i.e., 0.5) (Hanada et al. 2007; Ando 2009). It is worth noting that this factor $W_{\phi(0)}$ in the monaural autocorrelation function (ACF) is analogous to factor W_{IACC} in the interaural correlation function (IACF). $W_{\phi(0)}$ is deeply related to the characteristics of the frequency, depending on whether the components of the source signal are high or low.

4.2.2 Spatial Percepts of the Sound Field

Table 4.1 summarizes the auditory spatial percepts, which may be formulated by the spatial factors extracted from the IACF.

1. Localization in the Horizontal Plane

The perceived direction of a sound source in the horizontal plane can be expressed in terms of the spatial factors associated with the right hemisphere, which are extracted from the IACF, such that

$$L_{\text{Horizontal}} = S_{\text{R}} = f_{\text{R}}(\Phi_{\text{ll}}(0), \Phi_{\text{rr}}(0), \text{ IACC}, \tau_{\text{IACC}}, W_{\text{IACC}}) \tag{4.15}$$

where $\Phi_{\text{ll}}(0)$ and $\Phi_{\text{rr}}(0)$ signify sound energies of the signals arriving at the left and right ear entrances. It is well known that of the five spatial factors in Eq. (4.15), the most significant factors for the horizontal localization of a signal source are the interaural delay time, τ_{IACC}, and the interaural level difference, which is a function of $\Phi_{\text{ll}}(0)$ and $\Phi_{\text{rr}}(0)$ based on the difference between the sound energies at the two ears.

2. Apparent Source Width (ASW)

ASW has been formulated by the spatial factors extracted from the IACF, so that

$$\begin{aligned} S = S_{\text{R}} &= f(\text{IACC}) + f(\text{W}_{\text{IACC}}) + f(\text{LL}) \approx \alpha(\text{IACC})^{3/2} \\ &+ \beta(W_{\text{IACC}})^{1/2} + \gamma(\text{LL})^{3/2} \end{aligned} \tag{4.16}$$

where $\alpha \approx -1.64$, $\beta \approx 2.42$, $\gamma \approx 0.005$. Therefore, ASW increases with decreasing IACC and increasing W_{IACC} and LL.

3. Subjective Diffuseness

The scale values of subjective diffuseness, which is deeply related to the envelopment of sound, are inversely proportional to the IACC, so that

$$S = S_{\text{R}} \approx -\alpha(\text{IACC})^{3/2} \tag{4.17}$$

where $\alpha = 2.9$. The results of scale values obtained through the PCT together with values calculated using Eq. (4.17) as a function of the IACC are shown in Fig. 4.12 (Ando and Kurihara 1986).

4.3 Temporal and Spatial Percepts in Vision

The primary percepts of vision are similar to those of sound, consisting of temporal and spatial percepts (Ando 2009). As in auditory sensations, humans perceive the missing fundamental frequency of a visual flicker stimulus despite the absence of any frequency component at the fundamental. A typical temporal sensation of vision is thus the "pitch of the missing fundamental" of flickering light, which is described by τ_1 extracted from the temporal ACF. On the other hand, three salient spatial sensations of a two-dimensional "gray-level" texture, contrast, coarseness,

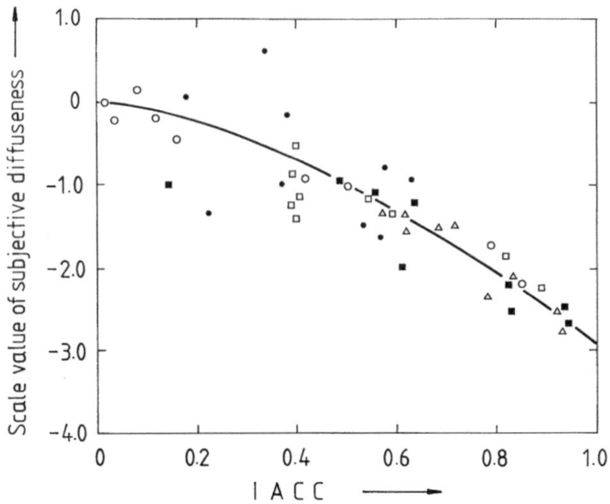

Fig. 4.12 Scale values of subjective diffuseness as a function of the IACC (calculated). Different symbols indicate different frequencies of the 1/3 octave bandpass noise: △: 250 Hz, ○: 500 Hz, □: 1 kHz, ●: 2 kHz, ■: 4 kHz. (____): Regression line by Eq. (4.17)

Table 4.3 Temporal and spatial sensations in vision in relation to factors extracted from the temporal ACF and spatial ACF, respectively

		Temporal sensations and spatial sensations			
Temporal or spatial ACF	Factors	Missing fundamental of flickering light	Contrast	Coarseness	Regularity
Temporal ACF					
	τ_1	X^a $(F = 1/\tau_1)$			
	ϕ_1	X (strength of the pitch)			
	τ_e	x^b			
Spatial ACF					
	$\Phi(0)$		$c_1\ \Phi(0)$	$c_6\Phi(0)$	
	δ_1		$c_2\ \delta_1$	$c_5\delta_1$	$c_4\delta_1$
	ϕ_1				$c_3\phi_1$
	δ_e				–

[a]X: Major factors to describe each sensation
[b]x: Attention to the fact that factors ϕ_1 and τ_e are mutually related

and regularity, are well described by the spatial factors extracted from the spatial ACF of monochrome visual patterns. Table 4.3 and Fig. 4.13 summarize the interconnections of the temporal and spatial sensations of the visual field, which are described in relation to factors extracted from the temporal and spatial ACF, respectively.

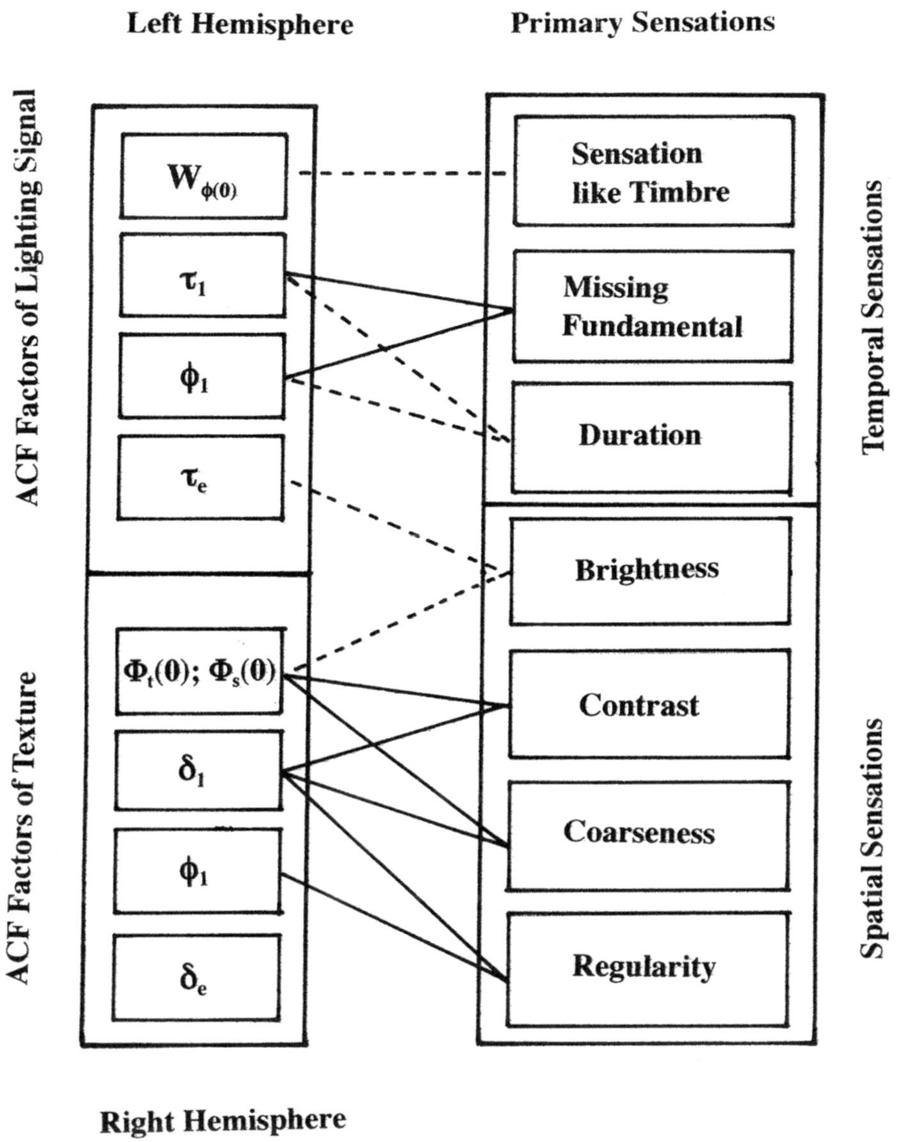

Fig. 4.13 Primary temporal and spatial sensations (percepts) of vision and the visual field associated with the left and the right cerebral hemispheres, respectively

4.3.1 Temporal Percepts of the Missing Fundamental Phenomenon in Vision

The "pitch" perceived at the fundamental frequency of visual complex signals of flickering light (i.e., the repetition rate of the flicker pattern) as shown in Fig. 4.14 is associated with the left hemisphere (L) and may be expressed in a manner similar to that of sound:

$$S = S_L = f_L(\tau_1) \approx 1/\tau_1[\mathrm{Hz}] \tag{4.16}$$

where the factor τ_1 is extracted from the temporal ACF. An example of the temporal ACF of the flicker stimuli for both tested conditions, the in-phase as well as the random-phase, is demonstrated in Fig. 4.15. The equation shown above has limitations per the range of the human visual system; therefore, it can reflect pitches under 3 Hz (Fig. 4.16) and frequency components under 40 Hz.

Remarkable results of the experiments clearly indicated that the most frequently perceived flicker rate for the complex waveforms corresponded to the missing fundamental component, which was not included in the power spectrum of the stimuli. The highest probability was seen at F0 = 2 and 2.5 Hz for the random-phase and in-phase conditions, respectively. Even though the periodicity of the waveform was unclear in the random-phase condition, the fundamental frequency was perceived, nonetheless (Fujii et al. 2000).

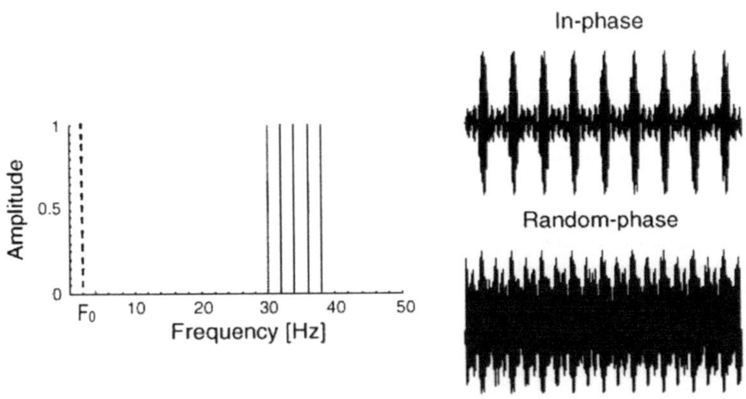

Fig. 4.14 An example of the spectrum of the complex signal used in the experiment. *Left*: Complex components are 30, 32, 34, 36, and 38 Hz, where the energy of the fundamental frequency ($F_0 = 2$ Hz) is absent. *Right*: Real waveforms in conditions of in-phase with remarkable peaks corresponding to the F_0 (*above*) and of random-phase (*below*)

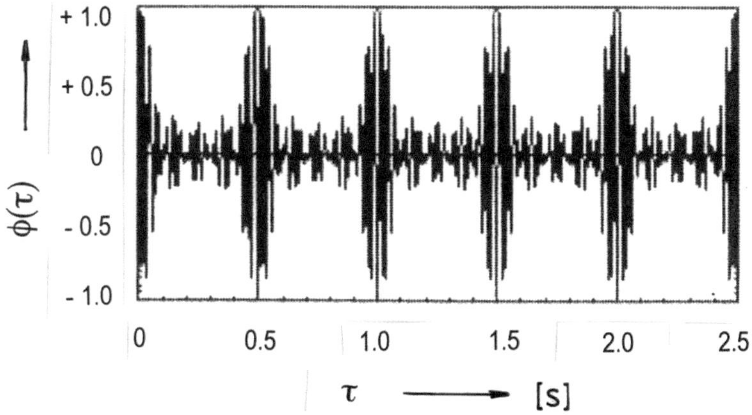

Fig. 4.15 An example of the temporal ACF of the stimuli of both conditions, in-phase and random-phase. The value of τ_1 corresponds to the fundamental frequency, which is not included in the power spectrum

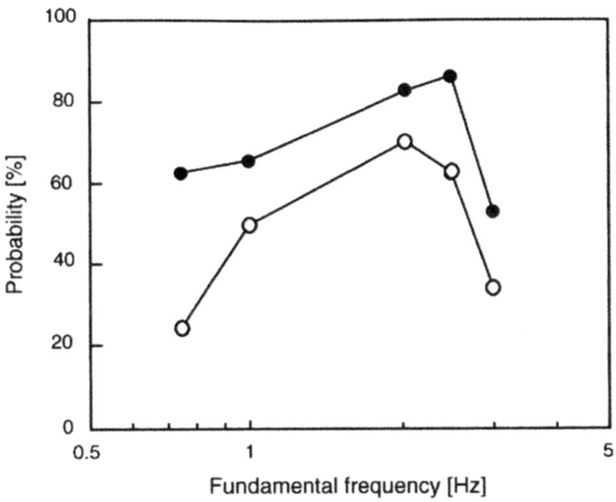

Fig. 4.16 The probability of response within $F_0(1 \pm 0.1)$ as a function of the fundamental frequency. *Filled circles* and *open circles* represent in-phase and random-phase conditions, respectively

4.3.2 *Spatial Percepts in Vision*

Materials applied for the texture experiments are shown in Fig. 4.17a (Brodatz 1966; Tamura et al. 1978; Fujii et al. 2003). Twelve botanical textures that were less harmonic (or regular) in structure (Fig. 4.17b), yet with a high degree of similarity

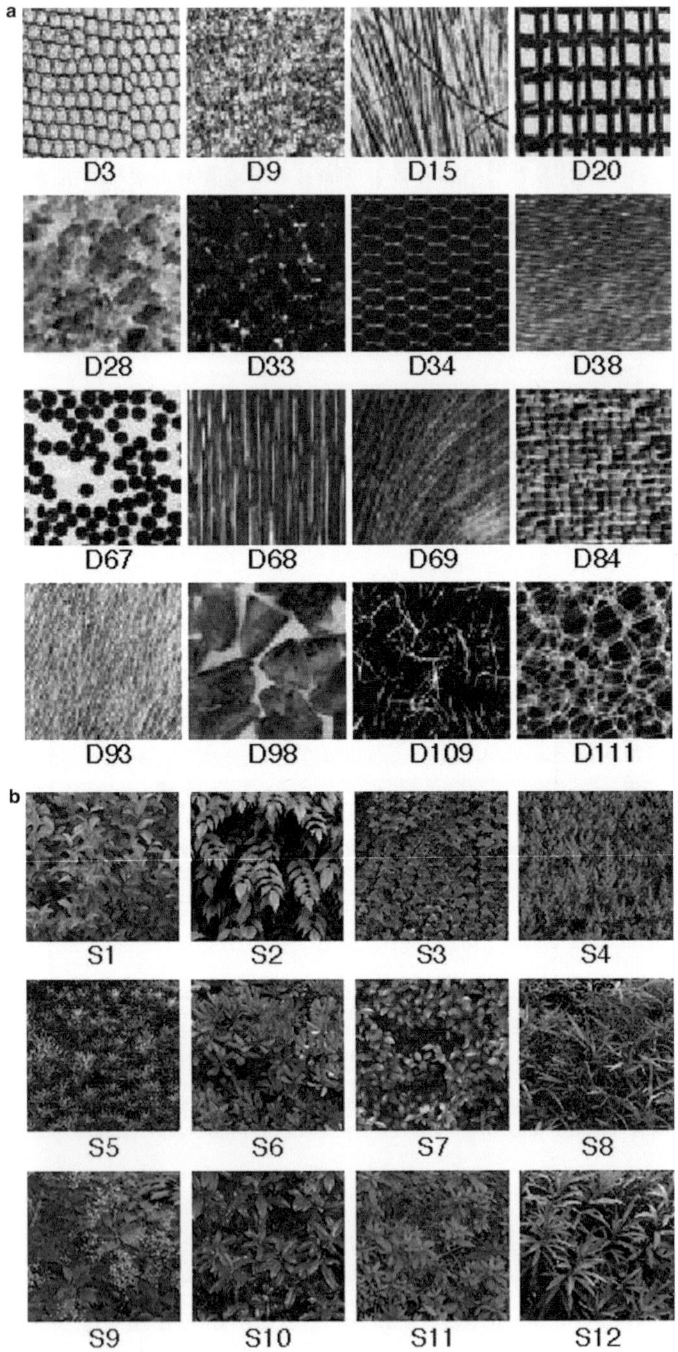

Fig. 4.17 (**a**) Texture set used by Tamura et al. 1978. (**b**) Botanical texture set applied for the additional experiment (Fujii et al. 2003)

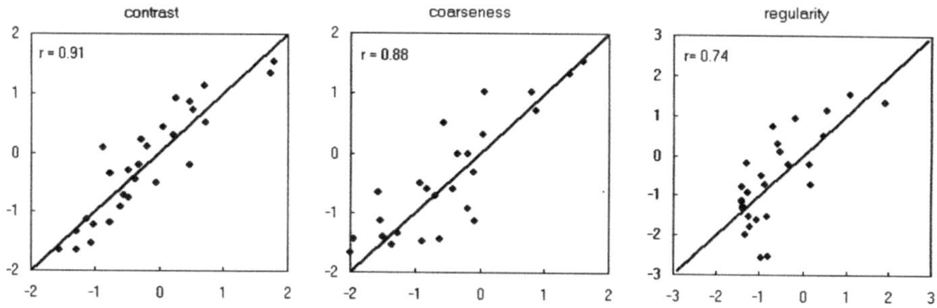

Fig. 4.18 Relationships between measured SV and those calculated by Eqs. (4.17, 4.18, and 4.19). (**a**) The SV of contrast ($r = 0.91$). (**b**) The SV of coarseness ($r = 0.74$). (**c**) The SV of regularity ($r = 0.88$)

between them, were used to discuss the spatial sensations in relation to multiple spatial factors extracted from the spatial ACF. Spatial percepts have been shown to deeply relate to and thus stimulate the right hemisphere. Spatial factors extracted from the two-dimensional spatial ACF are shown in Fig. 4.4. Symbol $\phi(\Delta x, \Delta y)$ signifies the spatial ACF for the spatial intervals Δx and Δy, ϕ_1 is the amplitude of the shortest major spatial interval (amplitude of the first maximum peak), $\delta_1$is the length of the shortest major spatial interval (the displacement of the first maximum peak in the ACF), and δ_e is the effective range of significant intervals (effective spatial duration of the ACF, which is defined as the displacement at which the envelope of the normalized ACF decayed at 0.1). By comparing the amplitudes and shortest spatial intervals in both directions, three factors were extracted from the ACF for the x direction only. Note that the x direction indicates more complicated spatial structures throughout the texture applied here.

According to the multiple regression analysis, the scale value (SV = S) of each spatial percept obtained by the PCT was approximately described by the spatial factors extracted from the ACF of the spatial signal of the monochrome "gray-level" visual pattern, so that

$$S_{\text{contrast}}\ c_1\Phi(0) + c_2\delta_1 \tag{4.17}$$

$$S_{\text{coarseness}}\ c_3\Phi(0) + c_4\delta_1 \tag{4.18}$$

$$S_{\text{regularity}}\ c_5\Phi_1 + c_6\delta_1 \tag{4.19}$$

where $c_1, c_2, \ldots c_6$ are coefficients. As a result of analyses of measured data, coefficients were obtained: $c_1 = 0.85$, $c_2 = 0.25$, $c_3 = 0.43$, $c_4 = 0.37$, $c_5 = 0.55$, and $c_6 = 0.74$. Figure 4.18 shows relationships between the measured SV (vertical axis) and the scale values (horizontal axis) calculated by Eqs. (4.17, 4.18, and 4.19). The correlation coefficients between the measured SV and calculated SV are $r = 0.91$ (contrast), 0.88 (coarseness), and 0.74 (regularity).

Thus far, we have clearly shown that the spatial factors extracted from the ACF analysis of monochrome (gray-level) visual patterns provide useful measures for representing the textural properties of contrast, coarseness, and regularity. Perceived contrast and coarseness are described by $\Phi(0)$ and δ_1, and perceived regularity is strongly related to the period δ_1 at the height of the peak ϕ_1.

References

Ando Y (1985) Concert hall acoustics. Springer, Heidelberg

Ando Y (1998) Architectural acoustics, blending sound sources, sound fields, and listeners. AIP Press/Springer, New York

Ando Y (2002) Correlation factors describing primary and spatial sensations of sound fields. J Sound Vib 258:405–417

Ando Y (2008) Theory of auditory temporal and spatial primary sensations. J Temporal Des Archit Environ 8:8–26. http://www.jtdweb.org/journal/

Ando Y (2009) Auditory and visual sensations. Springer, New York

Ando Y, Kurihara Y (1986) Nonlinear response in evaluating the subjective diffuseness of sound field. J Acoust Soc Am 80:833–836

Badcock DR, Derrington AM (1989) Detecting the displacement of spatial beats: no role for distortion products. Vision Res 29:731–739

Brodatz P (1966) Textures. Dover, New York

Damaske P, Ando Y (1972) Interaural cross-correlation for multichannel loudspeaker reproduction. Acustica 27:232–238

Fujii K, Kita S, Matsushima T, Ando Y (2000) The missing fundamental phenomenon in temporal vision. Psychol Res 64:149–154

Fujii K, Sugi S, Ando Y (2003) Textural properties corresponding to visual perception based on the correlation mechanism in the visual system. Psychol Res 67:197–208

Hammett ST, Smith AT (1994) Temporal beats in the human visual system. Vision Res 34: 2833–2840

Hanada K, Kawai K, Ando Y (2007) A study of the timbre of an electric guitar sound with distortion. In: Proceedings of the 3rd international symposium on temporal design, Guangzhou, November 2007, p. 96–99. J S China Univ Technol (Nat Sci Ed)

Henning GB, Herz BG, Broadbent DE (1975) Some experiments bearing on the hypothesis that the visual system analyzes spatial patterns in independent bands of spatial frequency. Vision Res 15:887–897

Inoue M, Ando Y, Taguti T (2001) The frequency range applicable to pitch identification based upon the auto-correlation function model. J Sound Vib 241:105–116

Mouri K, Akiyama K, Ando Y (2001) Preliminary study on recommended time duration of source signals to be analyzed, in relation to its effective duration of autocorrelation function. J Sound Vib 241:87–95

Saifuddin K, Matsushima T, Ando Y (2002) Duration sensation when listening to pure tone and complex tone. J Temporal Des Archit Environ 2:42–47. http://www.jtdweb.org/journal/

Sato S, Ando Y, Mellert V (2001) Cues for localization in the median plane extracted from the autocorrelation function. J Sound Vib 241:53–56

Sato S, Kitamura T, Ando Y (2002) Loudness of sharply (2068 dB/Octave) filtered noises in relation to the factors extracted from the autocorrelation function. J Sound Vib 250:47–52

Sayles M, Winter IM (2008) Reverberation challenges the temporal representation of the pitch of complex sounds. Neuron 58:789–801

Soeta Y, Maruo T, Ando Y (2004) Annoyance of bandpass filtered noises in relation to the factor extracted from autocorrelation function. J Acoust Soc Am 116:3275–3278

Sumioka T, Ando Y (1996) On the pitch identification of the complex tone by the autocorrelation function (ACF) model. J Acoust Soc Am 100(A):2720

Tamura H, Mori S, Yamawaki T (1978) Textural features corresponding to visual perception. IEEE Trans Syst Man Cybern SMC-8:460–472

Zwicker E, Flottorp G, Stevens SS (1957) Critical band width in loudness summation. J Acoust Soc Am 29:548–557

Chapter 5
Theory of Subjective Preference for Sound and Visual Fields

5.1 Theory of Subjective Preference for Sound Fields

Preferences as they present themselves to us – as the most primitive responses that steer organisms toward the direction of maintaining and propagating life – are always as relative as they are unique. In humans, preferences are deeply related to the sense of aesthetics. Therefore, making absolute judgments is problematic – needless to say that arriving at a result that questions its own objectivity is not what constitutes reliable information. Instead, preferences should be judged in a relative manner, such as is enabled by the paired-comparison test (PCT). It is the simplest and most accurate method, permitting both experienced and inexperienced persons starting from, for example, 2 years of age, to participate. The resulting scale values are appropriate for a wide range of applications. From the results of subjective preference studies with regard to the temporal and spatial factors of the sound field, the theory of subjective preference has emerged (Ando 1983, 1985, 1998, 2007).

Since the number of orthogonal acoustic factors, which are implicit in the sound signals arriving at both ears, is limited, the scale value of any one-dimensional subjective response may be expressed as a function of these factors:

$$S = g(x_1, x_2, \ldots x_I) \tag{5.1}$$

where $x_1, x_2, \ldots x_I$ are acoustic factors of the sound field and suffix I is the integer of the number of significant factors. As discussed in Chaps. 2 and 3, overall responses of the brain such as subjective preferences are described by both the temporal factors (processed in the left hemisphere) and the spatial factors (processed in the right hemisphere). Figure 4.6 shows the interconnections between the temporal and spatial factors and the corresponding primary auditory percepts, while Fig. 4.13 shows the same for primary visual percepts. It has been verified by a series of experiments that four objective acoustic factors act independently of the scale value, $I = 4$, when two of the four factors are varied simultaneously. Results

Y. Ando, *Brain-Grounded Theory of Temporal and Spatial Design*,
DOI 10.1007/978-4-431-55891-0_5

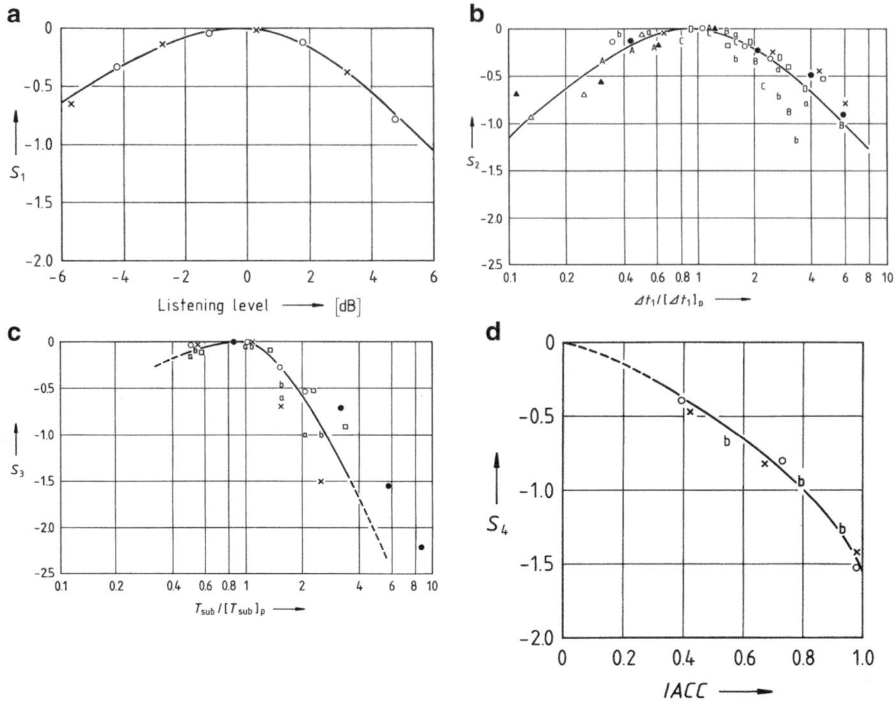

Fig. 5.1 Scale values of subjective preference obtained for simulated sound fields in an anechoic chamber, as a function of four normalized orthogonal factors of the sound field. *Different symbols* indicate scale values obtained from different source signals (Ando 1983, 1985). Even if different signals are used, a consistency of the scale value as a function of the normalized factor is observed fitting a single curve
(**a**) As a function of listening level *LL*. The most preferred listening level, $[\mathrm{LL}]_p = 0$ dB
(**b**) As a function of $x_2 = \Delta t_1/[\Delta t_1]_p$
(**c**) As a function of $x_3 = \mathrm{T_{sub}}/[\mathrm{T_{sub}}]_p$
(**d**) As a function of $x_4 = \mathrm{IACC}$

obtained from these subjective preference tests that used different source signals indicate that the four units of scale appear to be almost constant so that we may add individual scale values to obtain the total scale value for a sound field (Ando 1983),

$$\begin{aligned} S &= g(x_1) + g(x_2) + g(x_3) + g(x_4) \\ &= [S_2 + S_3]_{\text{Left Hemisphere}} + [S_1 + S_4]_{\text{Right Hemisphere}} \end{aligned} \tag{5.2}$$

where S_i, $i = 1, 2, 3, 4$ is the scale value obtained relative to each objective factor. Equation (5.2) indicates a four-dimensional continuity.

From the nature of the scale value, it is convenient to set its value to zero at the most preferred conditions without loss of any generality. Scale values of subjective preference obtained from other experimental series that used different music programs yield similar results when each factor is normalized by its most preferred value, as shown in Fig. 5.1a–d. The following common and approximate formula is given:

Table 5.1 Weighting coefficients α_i in Eq. (5.3) of the four orthogonal factors of sound fields obtained with a number of subjects

i	Normalized factors x_i	α_i	
		$x_i > 0$	$x_i < 0$
	Temporal factors associated with the left hemisphere		
2	$x_2 = \log(\Delta t_1/[\Delta t_1]_p)$	1.42	1.11
3	$x_3 = \log(T_{sub}/[T_{sub}]_p)$	0.45 + 0.75A	2.36 – 0.42A
	Spatial factors associated with the right hemisphere		
1	$x_1 = 20\log P - 20\log[p]_p$ (dB)	0.07	0.04
4	$x_4 = \text{IACC}$	1.45	–

$$S_i \approx -\alpha_i |x_i|^{3/2}, \quad i = 1, 2, 3, 4 \tag{5.3}$$

where x_i is the factor normalized by its most preferred value and the values of α_i are weighting coefficients as listed in Table 5.1. If α_i is close to zero, then a lesser contribution of the factor x_i on subjective preference is signified.

The factor x_1 is given by the sound pressure level difference, measured by the A-weighted network, so that

$$x_1 = 20\log P - 20\log[P]_p \tag{5.4}$$

P and $[P]_p$ being, respectively, the sound pressure or listening level (LL) present at a specific seat and the most preferred sound pressure that may be assumed at a particular seat position in the room under investigation. It is noteworthy that the binaural listening level (LL) can usually be approximated by a single microphone measurement. And

$$x_2 = \log\left(\Delta t_1 / \ [\Delta t_1]_p\right) \tag{5.5}$$

$$x_3 = \log\left(T_{sub} / \ [T_{sub}]_p\right) \tag{5.6}$$

$$x_4 = \text{IACC} \tag{5.7}$$

Thus, the scale values of preference have been formulated approximately in terms of the 3/2 power of the normalized objective parameters, expressed in terms of the logarithm of the normalized factors, x_1, x_2, and x_3. The remarkable fact is that the spatial binaural parameter $x_4 = \text{IACC}$ is expressed in terms of the 3/2 powers of its "real value," indicating a greater contribution than those of the temporal "logarithmic" parameters. As the nature of expression by Eq. (5.3), the scale values are not greatly changed in the neighborhood of the most preferred conditions but decrease rapidly outside this range. Since the experiments were conducted to find the optimal conditions, this theory remains valid in the range of preferred conditions tested for the four orthogonal factors.

This theory has been established based on evidence gathered by measuring neural activity in the central auditory signal processing system and the human cerebral hemispheres, which are deeply related and influential to the formation of scale values of subjective preferences for sound fields (Ando 2009). Under the preferred listening conditions, the brain repeats a similar rhythm in the alpha range over a longer period of time, and in addition, alpha rhythms persist and propagate over wider cortical territories. The most preferred conditions are discussed in Chap. 6.

5.2 Theory of Subjective Preference for Visual Fields

Similarly to the abovementioned preference theory for the sound field, the scale value of subjective preference for the visual field may be expressed in the form of Eq. (5.2) as well. As discussed in Chap. 3, the left and right cerebral hemispheres seem to be specialized for the temporal and spatial factors of the visual field, respectively. Therefore, independent influence between these two major sets of factors on the scale value of subjective preference is considered. Similarly to Eq. (5.2), the total scale value of subjective preference for a visual field could be formulated as

$$\begin{aligned} S &= g(x_1) + g(x_2) + g(x_3) + \cdots \\ &= \left[\sum S_i\right]_{\text{Left Hemisphere}} + \left[\sum S_j\right]_{\text{Right Hemisphere}}, \quad i = 1, 2, \ldots \text{ and } j = 1, 2, \ldots \end{aligned} \tag{5.8}$$

where ΣS_i represents the factors associated with the left hemisphere and ΣS_j represents the factors associated with the right hemisphere.

As listed in Table 5.2, the scale value of subjective preference might yield the common formula of Eq. (5.3), after obtaining the most preferred value of each temporal and spatial factor. Some examples of the most preferred temporal and spatial factors are discussed in Chap. 6.

Table 5.2 Weighting coefficients α_i in Eq. (5.3) of vision obtained with a number of subjects

x_i	α $x>0$	$x<0$
Temporal factors associated with the left hemisphere		
Flickering light, $x=\phi_1/[\phi_1]_p$	10.98	10.98
Moving single target		
1. Vertical movement, $x=T/[T]_p$	19.04	10.17
2. Horizontal movement, $x=T/[T]_p$	13.64	10.96
Spatial factors associated with the right hemisphere		
Texture, $x=\phi_1/[\phi_1]_p$	3.9	3.9

References

Ando Y (1983) Calculation of subjective preference at each seat in a concert hall. J Acoust Soc Am 74:873–887
Ando Y (1985) Concert hall acoustics. Springer, Heidelberg
Ando Y (1998) Architectural acoustics, blending sound sources, sound fields, and listeners. AIP Press/Springer, New York
Ando Y (2007 and 2014) Concert hall acoustics based on subjective preference theory. In Rossing TD (ed) Springer handbook of acoustics. Springer, New York
Ando Y (2009) In: Cariani P (ed) Auditory and visual sensations. Springer, New York

Chapter 6
Preferred Conditions of Temporal and Spatial Factors of Sound Fields

6.1 Preferred Conditions of Temporal Factors of the Sound Field

The two temporal factors are as follows:

1. Delay time of the first reflection after the direct sound Δt_1 (monaural and temporal factor), which is given by

$$\Delta t_1 = (d_1 - d_0)/c \ \ [\mathrm{s}] \tag{6.1}$$

where d_0 is the distance between the source position and the listener's position [m], d_1 is the path of the first or "the strongest reflection" arriving after the direct sound at the listener's position [m], and c is the speed of sound in the air [m/s].

2. Subsequent reverberation time after the first reflection T_{sub} (monaural and temporal factor), which is approximately given by Sabine's formula (Sabine 1900), such that

$$T_{\mathrm{sub}} \approx \frac{KV}{\overline{\alpha}S} \tag{6.2}$$

where K is a constant (about 0.162), V is the volume of the room [m^3], $\overline{\alpha}$ is the average absorption coefficient of the walls in the room, and S is the total surface of the room [m^2].

According to a systematic investigation that was built on simulating sound fields with multiple reflections and different reverberation times with the aid of a computer, and obtaining subjective preference values by the paired-comparison method, optimum design objectives and linear scale values of subjective preferences may be derived. The optimum design objectives can be expressed in terms of the subjectively preferred sound qualities described in more detail in this chapter,

Y. Ando, *Brain-Grounded Theory of Temporal and Spatial Design*,
DOI 10.1007/978-4-431-55891-0_6

which are related to the four orthogonal factors extracted from the autocorrelation function describing the sound signals arriving at the two ears. They clearly lead to a comprehensive set of criteria for achieving optimal design of concert halls as summarized below (Ando 1983, 1985, 1998, 2007).

6.1.1 *Optimal First Reflection Time After the Direct Sound (Δt_1)*

An approximation for the most preferred delay time has been expressed in terms of the minimum effective duration $(\tau_e)_{min}$ of the running ACF of the direct sound (source signal) and the total amplitude of reflections A, so that

$$[\Delta t_1]_p = \tau_p \qquad \approx (l - \log_{10} A)\,(\tau_e)_{min} \tag{6.3}$$

where the total pressure amplitude of reflections A is given by

$$A = \left[A_1^2 + A_2^2 + A_3^2 + \cdots\right]^{1/2} \tag{6.4}$$

where $A_n = d_0/d_n$ ($n = 1, 2 \ldots$) is the pressure amplitude of each reflection arriving at the listener's position and d_n is the path of the n-th reflection. The method of obtaining the value of τ_e is shown in Fig. 2.3.

The value of $(\tau_e)_{min}$ in a piece of music is generally observed in its most active part, the part with the least redundancy, the sharpest musical contrasts, and the one usually containing the most "artistic" expressive timing information such as *vibrato* or *accelerando* in the musical flow. For example, echo disturbances are easily perceived in musical segments where $(\tau_e)_{min}$ occurs. Even for a long music composition, musical flow can be divided into short segments, so that the minimal values of $(\tau_e)_{min}$ of the running ACF of the entire musical piece can be taken into consideration. This value is useful for matching musical programs to concert halls, for choosing which piece will sound best in a given concert hall. Methods for controlling the minimum value $(\tau_e)_{min}$ for vocal music performances have been discussed for this purpose (Ando 1998). If *vibrato* is introduced during singing, for example, it decreases $(\tau_e)_{min}$, thus blending the sound field with a short reverberation time (Kato et. al. 2007).

6.1.2 *Optimal Subsequent Reverberation Time After the Early Reflection (T_{sub})*

It has been observed that the most preferred condition of frequency response to reverberation time is just a flat without any frequency dependence. The preferred subsequent reverberation time, which is equivalent to that defined by Sabine, is expressed approximately by a constant multiple of the effective duration of the program material

$$[T_{\mathrm{sub}}]_p \approx 23(\tau_e)_{\min} \tag{6.5}$$

The total amplitude A of late reflections tested was in the range of 1.1–4.1, which covers the usual conditions of a sound field in a room, A being normalized by the pressure amplitude of the direct sound. Recommended reverberation times for different sound sources are given in Fig. 6.1. Furthermore, the design of a building must take into account its acoustical purpose:

1. A lecture and conference room should be designed for speech and an opera house mainly for vocal music but also for orchestral music.
2. For orchestral music, there may be two or three types of concert hall designs that best match the effective duration of the running ACF of the music programs that will be performed there. For example, Symphony No. 41 by Mozart, "Le Sacre du Printemps" by Stravinsky, and Arnold's Sinfonietta have short values of $(\tau_e)_{\min}$.
3. On the other hand, Symphony No. 4 by Brahms and symphony No. 7 by Bruckner are more typical of "slower" orchestral music.
4. Much longer values of $(\tau_e)_{\min}$ are usual for pipe organ music by Bach, for example.

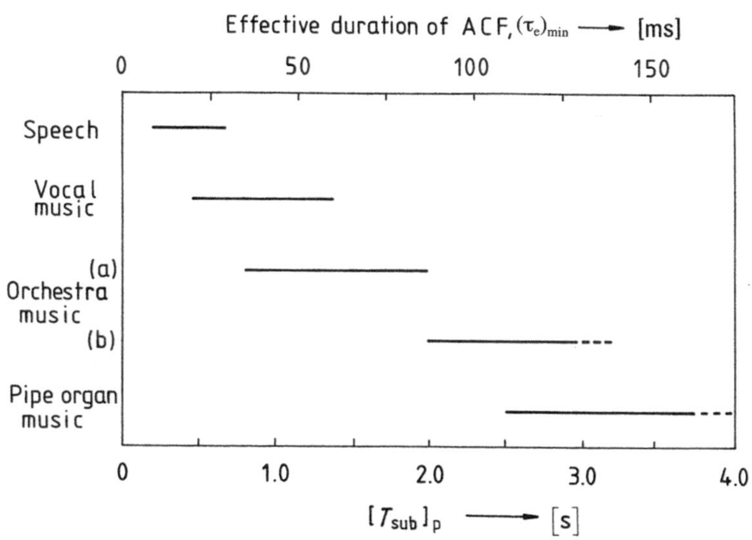

Fig. 6.1 Ranges of the most preferred reverberation time for speech and different music programs corresponding to the minimum value of the effective duration $(\tau_e)_{\min}$ that was extracted from the running ACF of source signals

Thus, the most preferred reverberation time for each sound source given by Eq. (6.5) can potentially play an important role in the selection of music motifs to be performed in a given concert hall.

6.2 Preferred Conditions of Spatial Factors of Sound Fields

6.2.1 Listening Level (LL)

The binaural listening level is the average sound-pressure level at the listener's ears. It is the primary factor that influences listening preferences for sound fields in concert halls. The preferred listening level depends upon the music and the particular passage being performed. The preferred levels obtained from 16 subjects were similar for two extreme music motifs, in the peak ranges of 77–80 dBA: 77–79 dBA for music motif A (the slow and somber Royal Pavane by Gibbons) and 79–80 dBA for music motif B (the fast and playful Sinfonietta by Arnold). As far as these two music motifs are concerned, about 79 dBA is a common value.

6.2.2 Optimal Magnitude of Interaural Crosscorrelation Function (IACC)

All available data derived from listeners with normal hearing ability indicated a negative correlation between the magnitude of the IACC and subjective preference, which reconfirms that dissimilarity between the signals arriving at the two ears is preferred (Ando 1998). Therefore, the preferred condition is given by

$$\mathrm{IACC} < 0.4 \tag{6.6}$$

This relation holds only under the condition that the maximum value of the IACF is maintained at the origin of the time delay, or near zero interaural delay, with the sound image directly in the front and an equal balance between the sound fields for the two ears, such that

$$\tau_{\mathrm{IACC}} = 0 \tag{6.7}$$

If balance of the sound field is not maintained for the two ears, then an image shift of the source may occur. To obtain a small magnitude of the IACC in the most effective manner, the directions from which the early reflections arrive at the listener should be kept within a certain range of angles from the median plane centered on $\pm 55°$. It is obvious that the sound arriving from the median plane $\pm 0°$ makes the IACC greater. Sound arriving from $\pm 90°$ in the horizontal plane is not

Table 6.1 Preferred conditions of temporal and spatial factors for the sound field

Four orthogonal factors	Preferred condition indicating the maximum value of subjective preference
Temporal factors associated with the left hemisphere	
Δt_1	$[\Delta t_1]_p \approx (1 - \log_{10}A)\ (\tau_e)_{min}$
T_{sub}	$[T_{sub}]_p \approx 23(\tau_e)_{min}$
Spatial factors associated with the right hemisphere	
LL = 20log[P] (dB)	About 80 dBA at a peak, but it depends on music and musical expression
IACC	IACC < 0.4

always advantageous, because the similar "detour" paths around the head to both ears cannot decrease the IACC effectively, particularly for frequency ranges higher than 500 Hz. For example, the most effective angles for the frequency ranges of 1 kHz and 2 kHz are roughly centered on ±55° and ±36°, respectively. To realize these conditions for different frequency components of source signals simultaneously, adopting geometrically uneven surfaces for side walls has been proposed (Ando and Kato 1976).

The most preferred conditions of the temporal and spatial factors of the sound field are summarized in Table 6.1.

References

Ando Y (1983) Calculation of subjective preference at each seat in a concert hall. J Acoust Soc Am 74:873–887

Ando Y (1985) Concert hall acoustics. Springer, Heidelberg

Ando Y (1998) Architectural acoustics, blending sound sources, sound fields, and listeners. AIP Press/Springer, New York

Ando Y (2007 and 2014) Concert hall acoustics based on subjective preference theory. In: Rossing TD (ed) Springer handbook of acoustics. Springer, New York

Ando Y (2009) Auditory and visual sensations. Springer, New York

Ando Y, Kato K (1976) Calculations on the sound reflection from periodically uneven surfaces of arbitrary profile. Acustica 35:321–329

Kato K, Fujii K, Hirawa T, Kawai K, Yano T, Ando Y (2007) Investigation of the relation between minimum effective duration of running autocorrelation function and operatic singing with different interpretation styles. Acta Acustica United Acustica 93:421–434

Sabine WC (1900) Reverberation. The American architect and the engineering record. Sabine WC Prefaced by Beranek LL (1992) Collected papers on acoustics, Peninsula Publishing, Los Altos, California, Chapter 1

Chapter 7
Subjective Preferences in Vision

7.1 Subjective Preferences for the Temporal Factor of Flickering Lights

In order to obtain basic data for the temporal design of visually perceived phenomena and environments, paired-comparison tests (PCT) were conducted to investigate subjective preference for a flickering light in terms of the ACF factors of the stimulus signal in the time domain (Soeta et al. 2002). It was found that the preferred sinusoidal period $[T]_p$ of the flickering light is roughly

$$[T]_p \sim 1.0 \text{ s}. \tag{7.1}$$

In order to obtain more representations of subjectively preferred conditions, a fluctuation was introduced to the preferred sinusoidal flickering light centered on 1 Hz (period of 1.0 s as given by Eq. (7.1)). In this procedure, the amplitude of the first maximum peak extracted from the temporal ACF of the flickering light stimulus, factor ϕ_1 shown in Fig. 7.1, was controlled, by changing the bandwidth of the noise (1, 2, 4, 8, and 16 Hz). In the natural environment, we perceive many visual aspects in temporal fluctuation, from twinkling stars to leaves in the wind and flow of water in a river.

As indicated in Table 7.1, the most preferred fluctuations of the flickering light for individual subjects as well as the averaged preference are described by the factor ϕ_1. Subjective preferences of individuals ranged $[\phi_1]_p = 0.27$–0.90. The averaged value for the participating subjects was $[\phi_1]_p \approx 0.46$ (Soeta et al. 2005). Remarkably, the resulting preferred fluctuation expressed by ϕ_1 is an intermediate of the values between those of the perfectly periodic sine wave ($[\phi_1]_p = 1.0$) and a perfectly random wave ($\phi_1 = 0$). The value of ϕ_1 corresponds to "pitch strength" in sound signals. In music performance, it is a kind of artistic expression in the temporal domain. It is possible to produce a visual light and even a music signal based on this temporal factor and also to blend visual light and sound.

Y. Ando, *Brain-Grounded Theory of Temporal and Spatial Design*,
DOI 10.1007/978-4-431-55891-0_7

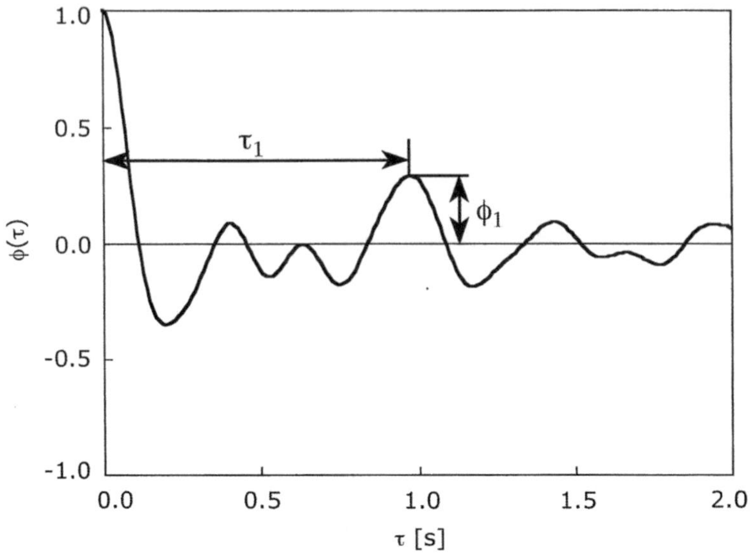

Fig. 7.1 Factors τ_1 and ϕ_1 extracted from the ACF of the flickering light

Table 7.1 The most preferred periods of the flickering light $[\phi_1]_p$ for each observer and the averaged value

Observer	$[\phi_1]_p$
A	0.51
B	0.50
C	0.47
D	0.58
E	0.45
F	0.90
G	0.27
H	0.33
I	0.33
J	0.31
Averaged	0.46

After obtaining the most preferred conditions for individual observers, the scale value of subjective preference may be obtained as a function of the normalized factor $\phi_1/[\phi_1]_p$ as shown in Fig. 7.2. Thus, the preference evaluation curve for both individual observers and for an average of all of the observers can be expressed as

$$S \approx -\alpha|x|^{3/2} \tag{7.2}$$

where $x = \log\phi_1 - \log[\phi_1]_p$. The averaged weighting coefficient has been obtained with a number of subjects, $\alpha \approx 11.0$.

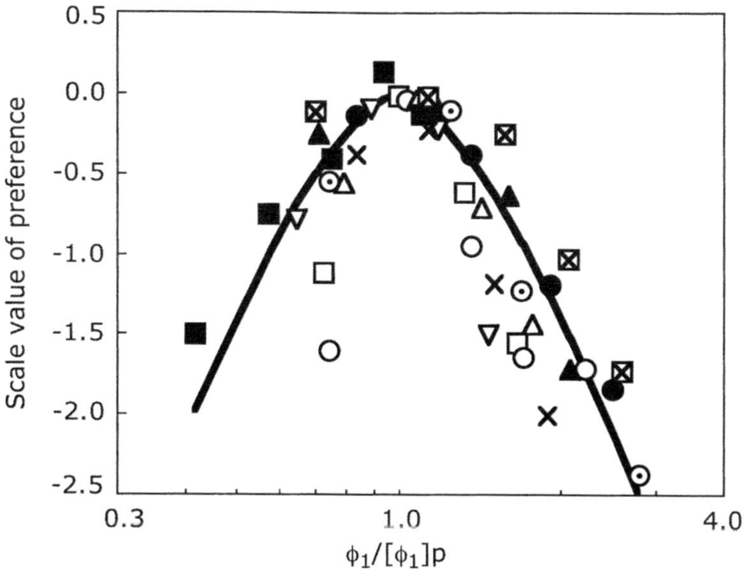

Fig. 7.2 The scale value of subjective preference as a function of the normalized factor $\phi_1/[\phi_1]_p$, where $[\phi_1]_p$ is the most preferred value of each observer as listed in Table 6.1

In the thermal environment, the effects of a "breeze" could be described by applying the factor ϕ_1, which is extracted from the ACF of wind speed. Readers are encouraged to embark on experiments in this field.

7.2 Subjective Preferences for the Temporal Factor of Oscillatory Movements

Results of preference judgments applying the PCT for sinusoidal movements of a single circular target without any fluctuation on a monitor screen are mentioned in this section. The period of stimulus movements was varied separately in the vertical or horizontal direction. Results indicate that the most preferred periods ($[T]_p$) for all subjects are about 1.0 s in the vertical direction and about 1.3 s in the horizontal direction. The curve of the scale values of preference may be commonly expressed by Eq. (7.1) with $x = \log_{10}T - \log_{10}[T]_p$ and $\beta = 3/2$.

The stimuli were displayed on a CRT monitor presenting 30 frames per second. Figure 3.2 shows the stimulus used in the experiment – a single white circular target moving sinusoidally (Soeta et al. 2003). The diameter of the target was subtended 1° of the visual angle (1.22 cm). The movement of the stimulus is expressed as:

Table 7.2 The most preferred periods $[T]_p$ of vertical and horizontal movements of the target for each subject and the averaged values

Subject	Vertical [s]	Horizontal [s]
A	1.15	1.28
B	1.05	1.82
C	0.78	1.31
D	1.16	1.79
E	0.85	0.91
F	0.83	1.05
G	1.08	1.31
H	0.81	1.04
I	0.93	0.98
J	1.10	1.13
Averaged	0.97	1.26

$$h(t) = A \cos(2\pi t/T) \tag{7.3}$$

where A is the amplitude and T is the period of the stimulus. In all experiments, the amplitude A was fixed at 0.61 cm on the monitor screen, corresponding to 0.5° of the visual angle. The white target and black background corresponded with gray levels 40 and 0.5 cd/m^2, respectively. The monitor presenting the stimuli was placed in a dark room 0.7 m away from the subject's eye position to maintain natural binocular vision.

Subjective preference for the period of movements in the horizontal and vertical directions was examined separately. The period of stimulus movement T in Eq. (7.3) was varied at six levels: $T = 0.6$, 0.8, 1.2, 1.6, 2.0, and 2.4 s. Thirty pairs combining six different periods constituted each series, and 10 series were conducted for all 10 subjects in the experiments by the PCT.

The most preferred period $[T]_p$ for each subject was estimated by fitting a suitable polynomial curve to a graph on which scale values were plotted as shown in Fig. 3.3. The individual values of preference for movement oscillation periods in the vertical and horizontal directions are listed in Table 7.2 and plotted in Fig. 7.3a, b, respectively, so that preference curves may be expressed by Eq. (7.2) also.

The global value of the most preferred period was about 1.0 s for vertical movement and about 1.3 s for horizontal movement as summarized in Table 7.3. Results from all subjects indicated that preferred periods in the vertical direction were shorter than in the horizontal direction ($p < 0.01$).

7.3 Subjective Preferences for Spatial Factors of Textures

The current section discusses whether or not the results pertaining to the abovementioned temporal fluctuation persevere in terms of the spatial fluctuations in texture (Ando 2009). Evaluation of subjective preference for textures is shown in

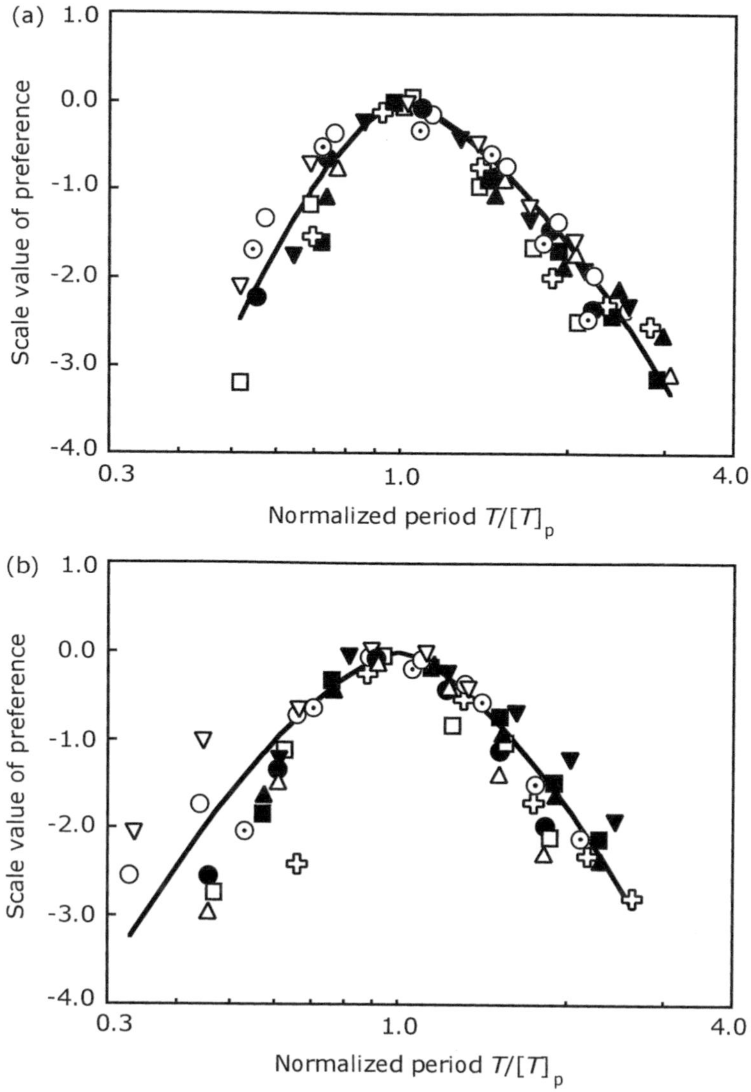

Fig. 7.3 (**a**) Normalized scale values of preference for individual subjects in the *vertical direction*. (**b**) Those in the *horizontal direction*. *Different symbols* indicate the scale values obtained by different subjects

Table 7.3 Preferred conditions of temporal factors for vision with different movements

Temporal	Factors	Flickering light	Moving single target
ACF	τ_1	$[\tau_1]_p \approx 1.0$ s	$[\tau_1]_p \approx 1.0$ s (vertical)
			$[\tau_1]_p \approx 1.3$ s (horizontal)
	ϕ_1	$[\phi_1]_p \approx 0.46$	–
	τ_e	–[a]	–

[a]–: Factors to be examined for their influence/insignificance on the respective attribute

Fig. 7.4 Two-dimensional spatial textures used with the value of ϕ_1 for the subjective preference judgment

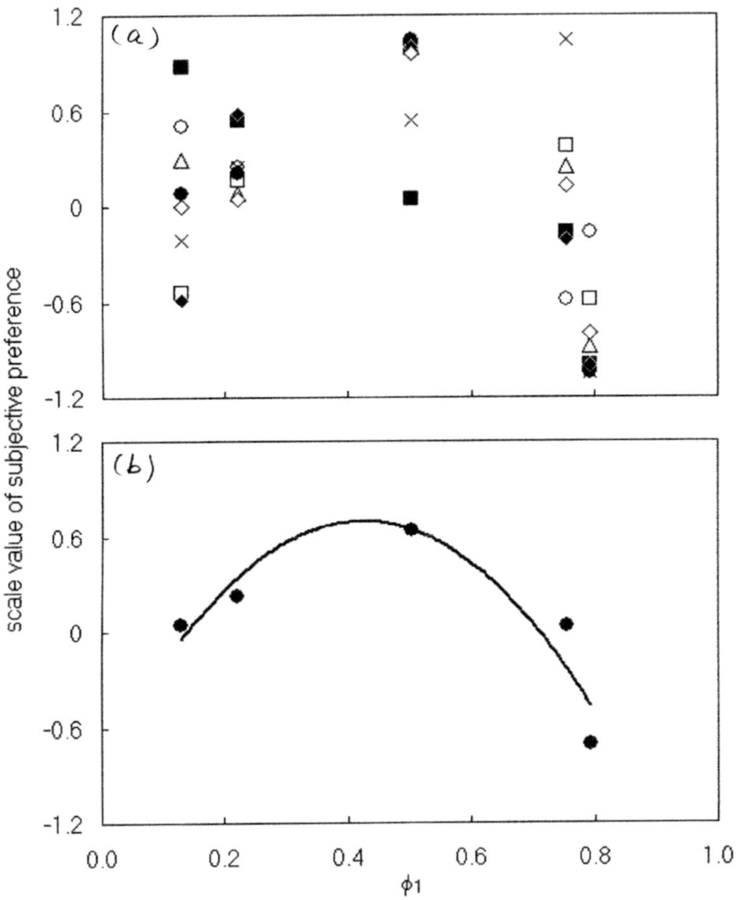

Fig. 7.5 (**a**) Scale values of preference for individual subjects. (**b**) Averaged preference values and a fitting curve with the 3/2 power of ϕ_1 in Eq. (6.1)

Fig. 7.4, which was conducted by the PCT by changing the value of factor ϕ_1 for perceived texture regularity. Results of subjective preference SV for all ten subjects are shown in Fig. 7.5. The upper figure shows a great degree of individual differences. By averaging the scale values of all subjects, the most preferred value for texture regularity was found at $[\phi_1]_p \approx 0.41$. The weighting coefficient α in Eq. (7.2) with all ten subjects tested was about 3.9. It is worth noting that the most preferred subjective preference for the flickering light with fluctuation as described in Sect. 7.1 was $[\phi_1]_p \approx 0.46$ as well. Thus, a certain degree of fluctuation in both temporal and spatial factors is a visual property affecting subjective preference.

As discussed in Sect. 7.1, subjective preference for a flickering light was obtained by the PCT, which was investigated in terms of the ACF factors of the signal in the time domain. The most preferred degree of fluctuation was found at $[\phi_1]_p \approx 0.46$, and the scale value of preference is formulated approximately in terms of the 3/2 power of the normalized ϕ_1 of a flickering light by the most preferred value, $[\phi_1]_p$ similar to Eq. (5.3).

In the natural environment, there are many visual aspects that we perceive in temporal fluctuation, such as leaves in the wind and clouds in the sky, twinkling stars due to air currents, and flowing water in a river. Flames in a bonfire and glitters of sunlight reflected off the surface of water provide us with a lively and splendid environment. Such phenomena may stimulate the left hemisphere as shown in Table 3.1.

In addition, preference judgments using the PCT for sinusoidal movements of a single circular target without any fluctuation on a monitor screen were performed. The period of stimulus movements was varied separately in the vertical or horizontal direction. Results showed that the most preferred periods ($[T]_p$) for all subjects were about 1 s in the vertical direction and about 1.3 s in the horizontal direction. As seen from Fig. 7.5b, the curves of the scale values of preference may be commonly expressed by Eq. (5.3) with $x = \log_{10}T - \log_{10}[T]_p$ and $\beta = 3/2$ (Soeta et al. 2003).

As an application of this theory, artistic expressions of color modulation and sequential form in a drawing, which may stimulate the left hemisphere, are discussed in Sect. 10.5.

References

Ando Y (2009) Auditory and visual sensations. Springer, New York

Soeta Y, Uetani S, Ando Y (2002) Relationship between subjective preference and alpha wave activity in relation to temporal frequency and mean luminance of a flickering light. J Opt Soc Am A 19:289–294

Soeta Y, Ohtori K, Ando Y (2003) Subjective preference for movements of a visual circular stimulus: a case of sinusoidal movement in vertical and horizontal directions. J Temporal Des Architect Environ 3:70–76. http://www.jtdweb.org/journal/

Soeta Y, Mizuma K, Okamoto Y, Ando Y (2005) Effects of the degree of fluctuation on subjective preference for a 1 Hz flickering light. Perception 34:587–593

Chapter 8
General Theory for Designing Physical Environments Incorporating Spatial and Temporal Factors

8.1 Biological Rhythms and Discrete Periods in the Physical Environment

In our life, as shown in Fig. 8.1, experiences of discrete temporal periods of the human and physical environments can be blended. Remarkably, there are certain significant periodic eigenvalues, in both human biological rhythms and physical environmental activities in the time domain, to be synthesized. The shortest period is related to the minimum firing interval of about 0.8 ms that is related to pitch perception. The instantaneous periods of 0.5–5 s that correspond to the psychological present are associated with brain waves, pulsation, breathing, and, therefore, also with the perception of speech, music, and even the twinkling stars, as well as the susurrus from the leaves and the glitter of a rippling water surface. Following such periods, there are longer eigenvalues. About 90 min corresponds to the period of rapid eye movement (REM), which is part of the basic rest-activity cycle that continues throughout day and night, occurring between 10 and 20 times a day (Othmer et al. 1969; Kripke 1972), depending on the length of the cycle (70–150 min). The basic rest-activity cycle is deeply related to one session of a concert, lecture, or any other type of work activity that requires concentrated thought. There is a deep connection between the circadian rhythm and the amount of sunlight, that is, the Earth's rotation period and daily human activity. The week that is formed by social custom for work and leisure is associated with, for example, the planning period of concert and drama, or any other type of social activity. The next temporal eigenperiod is concerned with the movement of the Moon that is deeply related to moonlight opera and other culturally relevant festivities and ceremonies. The revolution of the Earth around our Sun produces changes associated with the four seasons, like the changing color of the leaves, increasing and decreasing amounts of precipitation, and so on, as well as annual festivities such as the New Year opera. The black spots on the Sun which appear about every 11 years may influence environmental conditions on the Earth to a greater or lesser degree

Y. Ando, *Brain-Grounded Theory of Temporal and Spatial Design*,
DOI 10.1007/978-4-431-55891-0_8

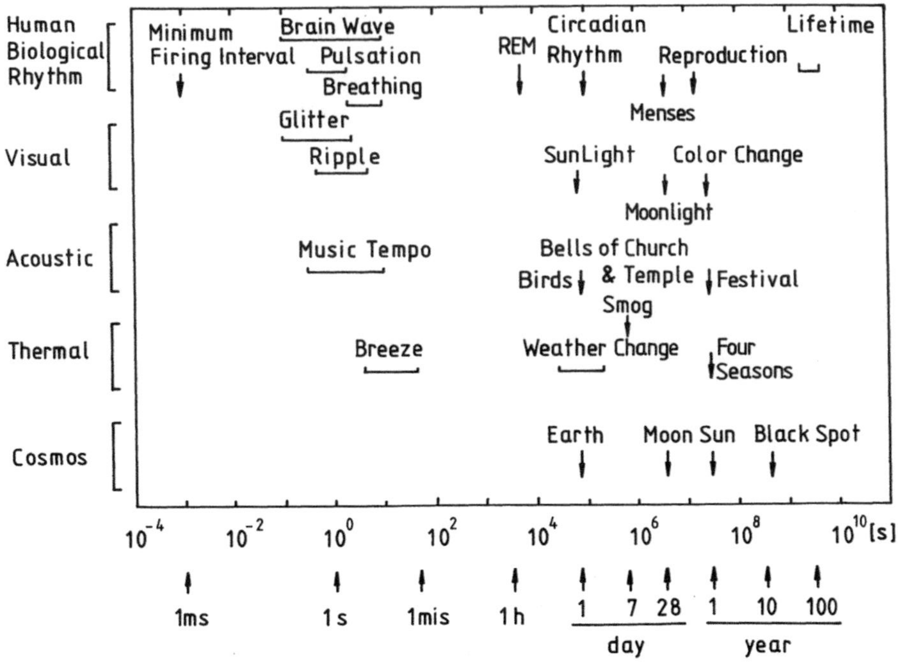

Fig. 8.1 Human rhythms and discrete periods of the natural physical environment to be considered in temporal design

(e.g., it has been assumed that there could be a link between solar activity and earthquakes). The alternation of generations approximately every 30 years, and the human life span of about 90 years, should be considered in the planning of houses, as the development of individual personality is in accordance with the schedule of life. There are many other cosmic oscillations that occur over much longer temporal periods such as the Milankovitch cycles (21,000... 41,000 years) which describe changes in the Earth's orbit around the Sun, manifesting in climatical changes due to the changing amount of solar energy reaching the Earth. The longest period of the universe is in the order of 10^{12} years, corresponding to the period from one big bang to the other (Appendix A3).

Thus, a most crucial factor to consider in the temporal dimension of the environment is what can be conceptualized as cycles, so that we do not need to consider "real-time" periods with possible scalar infinities. Every aspect of the passage of time is bound with cycles: birth and death, the changing of the seasons, sleeping and waking, work and leisure.

The present theory suggests that these cycles be explicitly recognized during the design process. The passage of time in the designed environment should be as consciously considered as the three-dimensional organization of the space itself.

8.2 Theory of Designing Physical Environments

It is assumed that Eq. (5.2) for calculating the scale value of subjective preference for the sound field is held for all physical environments including auditory, visual, thermal, and vibratory fields, such that

$$S = [g_L(x) + g_R(x)]_{\text{Sound field}} + [g_L(x) + g_R(x)_{\text{Visual field}} + g_L(x) + g_R(x)]_{\text{Thermal field}} + [g_L(x) + g_R(x)]_{\text{Vibratory Field}} \tag{8.1}$$

where $g_L(x)$ and $g_R(x)$ are the temporal and spatial factors for each physical field associated with the left and right hemispheres, respectively. For the temporal and spatial factors of thermal and vibratory environments, much is not known regarding the specialization of the hemispheres; however, evidence indicates that these factors are supposedly processed by different areas in the brain. Equation (8.1) holds at least in the neighborhood of the optimal conditions for each physical factor, avoiding extreme physical environments like, for example, listening to music in a temperature below 0 °C or in an intense vibratory and noisy environment such as in a small motor boat. Table 8.1 lists the significant factors to consider when designing physical environments.

As shown in Fig. 8.1, the crucial factor in the temporal dimension of the environment is the periodic cycle. The present theory is built upon both the temporal and spatial factors in each type of physical environment. In particular, such discrete periods should be explicitly recognized during the design process of any human environment. The passage of time in the designed environment should be as consciously considered as the three-dimensional organization of the space itself.

Human experience of temporal duration is not linear as physical time is, because it differs according to the activity a person is engaged in as discussed in Appendices A1 and A2. In spatial design, there are specific and characteristic dimensions of space by which its purpose and content are outlined, such as a room, a house, a building, a region, an urban site, and so on. Thus, the offered matrix consists of elements characteristic to both time and space.

8.3 Interaction Between Human Development and Environment

Let us delve deeper into the idea that each individual is born into this world with a specific message or mission. Our individuality is expressed already at the genetic level, and the farther a person evolves, the more crystallized and amplified this uniqueness becomes by experience and the conditions set forth by the environment. As the saying goes, no two snowflakes are alike, and the same applies to the infinite

Table 8.1 Physical environmental factors to be designed

Physical environment	Spatial factors	Temporal factors
Sound field	1. Listening level, LL	2. Initial time delay gap between the direct sound and the first reflection, Δt_1
	4. Interaural crosscorrelation, IACC	3. Subsequent reverberation time, T_{sub}
Visual field	1. Lighting level	2. Properties of movement function of reflective surface, T
	3. Properties of the reflecting surface	
	4. Spatial perception including distance factor	
Thermal field (assumed)	1. Temperature and humidity sensors distributed throughout the body	2. Relative humidity
		3. Temperature
		4. Air movement (e.g., breeze)
Vibration field (assumed)	1. Amplitude	2. Temporal factors extracted from the impulse response between the source position and observation position
	3. Vibration sensors distributed throughout the body	

Note: In addition to these variables, characteristics of sources in terms of the factors extracted from the temporal ACF, location change of sources, and observer's activity should also be taken into consideration

array of individuals that have passed and will pass throughout space and time. It is true, therefore, that every person has the potential of a special and unique ability to discover a part of the unknown as shown in Figs. P.4 and P.5. Such individual discoveries and missions, however, sometimes remain invisible not only to the public eye but also to the individuals themselves.

As shown in Fig. 8.2, the dimension of the head of unborn babies is relatively large in the womb compared to the rest of the body. This fact suggests that environment (including its temporal aspects) plays an important role in the development of the human brain after birth as well. For instance, there are particular temporal windows during the development of a human being where the ability to absorb different types of knowledge is at its peak. Such specific timeframes have been discovered for the benefit of optimizing the learning of languages, music, and mathematics, for example. Such pivotal early environments may also express themselves as a skyline patterned by a mountain ridge or a flowing river near one's native place. Before going to kindergarten and elementary school, the neural systems in the brain are almost fully developed, with deep marks embedded from the nearest (natural) environment.

One of the most significant and closest environments to a newborn baby is his/her room in the house. Designers and architects should recognize that such environments

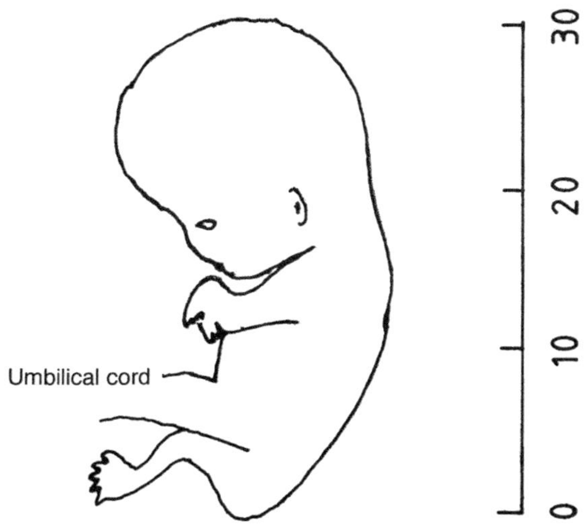

Fig. 8.2 Sketch of unborn baby at 52 days. The head is relatively large compared to other parts of the body, which indicates early development of the brain part. This suggests environmental design at the initial stage of human life relating to the brain is extremely important for the development of personality

act as nonverbal "teachers" for the newborn in the early stage of life. Therefore, it is highly recommended that during the design phase of houses and general urban planning, the relation between the facility and the human brain be carefully considered. In a house, for example, the existence of an area for clay work and painting is likely to play an important role in the development of the right hemisphere. In towns, facilities such as museums, concert halls, libraries, churches, and other institutions are most likely to influence the development of the third stage of life. Just as the limbs of the fetus develop later than the head, highways and communication systems in urban planning can be developed later on as well. Otherwise, the resulting image is that of new settlements where young parents have moved with their children – towns that become obsolete within 30 years simply because lack of cultural facilities does not attract new people to move there or even visit.

Cosmic weather influences our global climate and change is a natural part of life. In view of these changes, a more dynamic approach to design, one that incorporates temporal design of the environment, should be considered. Realization of this concept is not easy unless new values in addition to economics are introduced. Such values are the building blocks of human life, namely, the well-being of the body and the mind. For this purpose, an additional value of the third stage of life has been defined in the Preface section of this volume. Due to deep interaction between our cerebral hemispheres and the temporal and spatial aspects of the different physical fields that we perceive as shown in Fig. P.1, environments should be thoroughly designed. Such environments, when realized, greatly influence the development of both hemispheres of our brains.

8.4 House Environment as a Nonverbal Teacher for Newborns

Newborn babies and young children are very interested in the surrounding environment, always either touching and tasting everything they come into contact with, or asking, "What is this?" and "What is that?" They learn quickly through temporal and spatial physical environments inside the house and outside. This early time in life plays an indubitably important role in the development of the personality, which is deeply rooted in the left and right cerebral hemispheres.

There are many exciting environments for newborns, for example:

1. The mother's voice when speaking and singing without watching her stimulates the left hemisphere and thus its development.
2. Music programs that are widely selected (not just nursery songs, but classical music and swinging jazz) trigger the developing person's preferences related to the personality. These can be easily tracked from the child's behavior when listening to music that activates mainly the left hemisphere (Sect. 2.1).
3. A pendulum clock moving at a period of about 1.0 s signifies a room that is not inanimate but very much alive, while also activating the left hemisphere (Sect. 7.2).
4. Intermittent fresh air breezes produce gentle sounds from tree leaves and activate the left hemisphere.
5. Pieces of furniture, walls, ceilings, and floors: Well-designed furniture pieces, ceilings and floors, as well as walls with drawings and artistic works support the development of the personality that emerges from both cerebral hemispheres.
6. Windows looking outside, natural light, gardens with flowers, leaves on trees moving in the wind, water surfaces of small ponds, the Sun, the Moon, twinkling stars, clouds, rainfall, people in movement – all such phenomena constitute a wonderful environment where both hemispheres of the brain can flourish.
7. A bathroom with warm and cool water acts as a toy, stimulating both hemispheres of the brain as well as keeping the body clean.
8. Flames from a fireplace produce a wonderful sight in addition to providing warm satisfaction in a living room.

References

Othmer E, Hayden MP, Segelbaum R (1969) Encephalic cycles during sleep and wakefulness in humans: a 24-hour pattern. Science 164(3878):447–449

Kripke DF (1972) An ultradian biological rhythm associated with perceptual deprivation and REM sleep. Psychosom Med 3:221–234

Chapter 9
Design Procedures for Matrix Elements Consisting of Discrete Periods and Spatial Ranks

9.1 Acoustic Environment

It has become a matter of fact that among the different physical environments (e.g., acoustic, visual, thermal), prevalent daily noise affects the development of the human brain and, therefore, the well-being of the body and mind. The need for meticulous planning of the acoustical environment cannot be overemphasized and should be a priority in the design process. Sound environments also influence our experience of temporal duration, i.e., we feel time passing either faster or more slowly depending on the quality of the temporal design (or lack thereof), and this subjective experience is much more important than physical time itself (Appendix A1; Ando 1977; Ando and Ando 2010). The selected elements numbered in the matrix shown in Fig. 9.1 are proposed for consideration in the design process.

(1–3) The indication of 1 s in the horizontal axis of the figure signifies every moment that is related to the range of the "psychological present" (Fraisse 1984). This may deeply relate to the sound signal duration to be analyzed corresponding to the auditory temporal window, which is approximately given by 30 times of $(\tau_e)_{min}$ (Mouri et al. 2001). The spatial shapes of an opera house and a concert hall may be determined by minimization of the IACC at each seat, which is the representative spatial factor related to the space form of the entire building (Maki 1997; Ikeda 1997). A well-designed concert hall can act as a second musical instrument expressing both the temporal and spatial sensations of music (Ando 2007, 2009). A sound reproduction system, for instance, is designed by use of both temporal and spatial factors of the sound field (see Appendix II in Ando 1998).

(4–5) Since the brain's concentration is limited to about 90 min which is directly related to rapid eye movement (REM) sleep (Sect. 8.1), an intermission in opera and concert programs is an effective tool to employ when designing optimal conditions. A refreshment corner serving wine and soft drinks in a well-designed

Y. Ando, *Brain-Grounded Theory of Temporal and Spatial Design*,
DOI 10.1007/978-4-431-55891-0_9

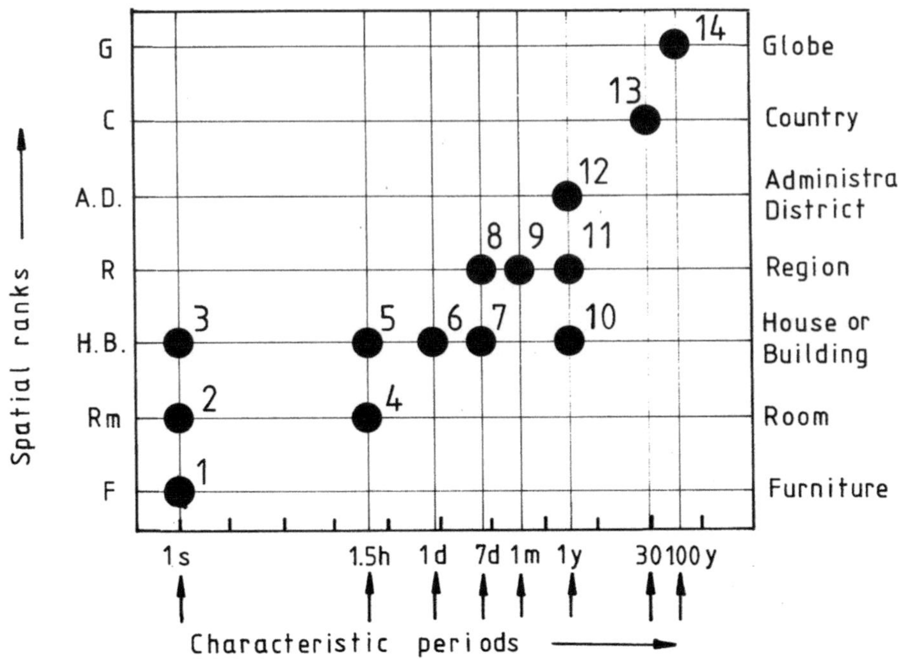

Fig. 9.1 Selected elements in the matrix with discrete periods and distinct ranks of space to be considered in acoustic environmental design. Numbers *1* through *9* indicate certain temporal and spatial design projects. A tendency is observed that small spaces are designed mainly for short terms, and large spaces are designed mainly for long terms

lobby serves as a meeting place for members of the audience as well as performers for discussing and exchanging opinions.

(6) Planning the flow throughout the whole building including the rehearsal room and the storage of instruments is considered based on daily schedule.

(7–9) Running concert programs based on weekly and/or monthly cycles may help to distribute notice without additional cost for each independent opera and concert program.

(10) Annual maintenance of the building or house is considered at the design stage during selection of materials.

(11–12) The design process should include the changing of the four seasons as well as spatial planning of the soundscape. This includes birdsong, susurring sounds from the leaves, and sounds of flowing water. Annual environmental noise in urban planning should be executed by minimizing long-term effects on unborn babies and children as discussed in Sect. 14.3 (Ando and Hattori 1977a, b).

(12–14) Urban planning should foresee and minimize the long-term effects of exposure to years of environmental noise from jet planes, motorcars, and trains, all of which affects the development of unborn babies and the specialization of their cerebral hemispheres (Ando et al. 1975; Ando 1988, 2001) as well as the development of height in children (Schell and Ando 1991). It has been shown that people who live surrounded by environmental noise feel time passes faster (Ando 1977) so that they always feel busy without performing any particular job. These effects of exposure to daily noise accumulate over several months or years, even in environments where the period of 24 h with a quiet night is the accepted norm. The effects that exposure to continuous noise has on the development of children remain present over the entire generation.

9.2 Visual Environment

Figure 9.2 indicates selected elements to be considered in the design of visual environments.

(1–3) Movement of tree leaves, scattering of sunlight in the air after reflection from water surfaces, streams of small rivers, twinkling stars, and many other natural phenomena stimulate the left cerebral hemisphere (Sect. 3.3). Their effect is related to the effective duration of the autocorrelation function (ACF) of

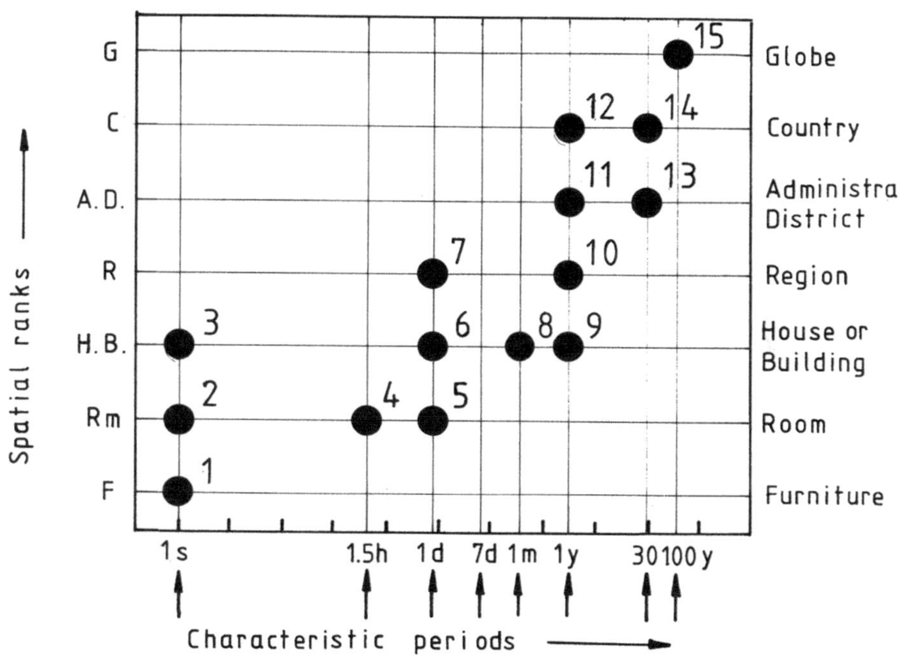

Fig. 9.2 Selected elements in the matrix with discrete periods and distinct ranks of space to be considered in visual environmental design. Numbers *1* through *9* indicate certain temporal and spatial design projects. A tendency is observed that small spaces are designed mainly for short terms, and large spaces are designed mainly for long terms

auditory and visual signals, τ_e, which is a unit of the temporal window as discussed earlier in this volume. Spatial sensations such as those derived from, for example, artistic drawings and photographs on the wall, stimulate the right cerebral hemisphere (Sperry 1974; Levy and Trevarthen 1976).

(4) Lighting as part of the design of a home environment can be controlled in the study, kitchen, living room, and bedroom. Lighting can be designed with regard to periods or sessions of 30–90 min due to REM. Blue and cool lights activate the brain in the study and the kitchen, while orange and warm lights are useful for making humans relax in the living room and the bedroom.

(5–6) As for spatial design on walls and floors, a certain degree of fluctuation in periodic textures will maximize subjective preference (Sects. 7.3 and 10.5). Natural light from the Sun, Moon, and stars can be introduced into rooms in a house through carefully planned windows (Bosworth and Ando 2006). Windows that enable experiencing the change of natural light or colors in the garden are considered superior for the benefit of the brain. Lateral lights are the most effective for looking at a human face.

(7) Duration times for areas of sunshine or shadow in a region can be planned as usual. The method for calculating sunshine and shadows from mountains and buildings is well known as a function of daytime.

(8) The changing Moon can be observed over the course of a month at well-designed viewing spots of houses and buildings, as in traditional Japanese houses such as, for example, Katsura Imperial Villa and Shisendo in Kyoto.

(9–12) Gardens, public parks, and walkways as well as plantations are designed according to the four seasons. Taking care of plantations may help the development of the mind as well as the brain and body for incurring further creations.

(13–15) Particularly, plantations in and close to cities play an important role in realizing good vision. They also work in favor of the third stage of life by offering “soil” for ideas to stem from by stimulating the act of contemplation of nature. Policies for the long-term design of the environment, which influence on people crosses a time span of 30 or more years and affects the health of humans either directly or indirectly, should be conscientiously planned by local governments, the government, and/or the United Nations.

9.3 Thermal Environment and Energy Conservation

Elements of the thermal environment to be designed are shown in Fig. 9.3. Some of the elements have been previously discussed, while others are not so well known.

(1–2) Design criteria for the thermal environment of rooms including a creative work space (CWS) and a creative kitchen space (CKS) consist of both temporal and spatial factors. A typical expression of a temporal factor is a breeze, which is defined by the fluctuation of wind velocity as a function of time. It can be expressed by the temporal factor ϕ_1 extracted from the ACF of wind velocity,

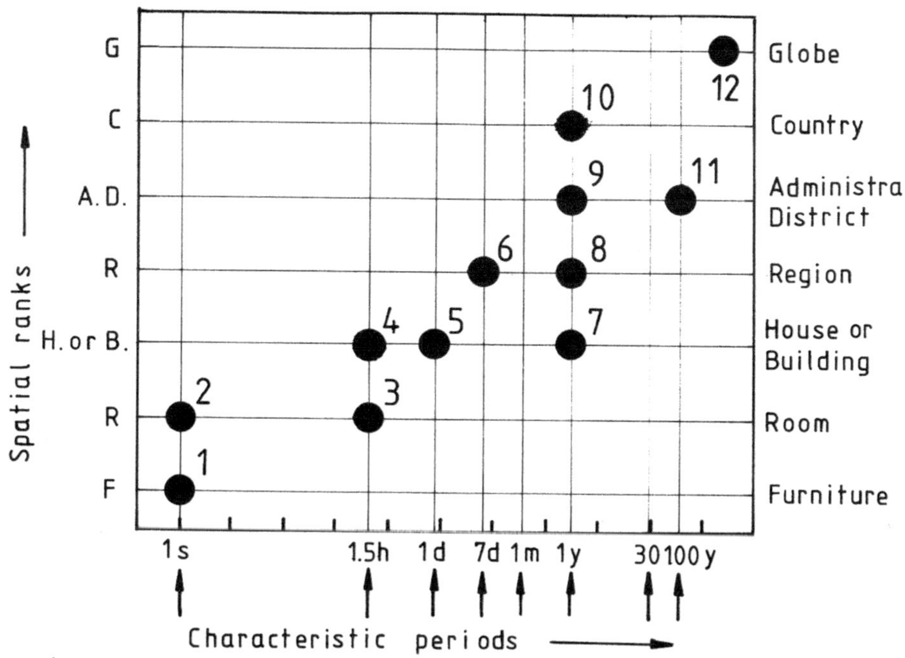

Fig. 9.3 Selected elements in the matrix with discrete periods and distinct ranks of space to be considered in thermal environmental design. A tendency is observed that small spaces are designed mainly for short terms, and large spaces are designed mainly for long terms

in a manner similar to the flickering light with fluctuation (Sect. 7.1: Soeta et al. 2002a, b), where the most preferred condition $[\phi_1]_p$ is assumed to be centered around 0.4. The most effective spatial factor to consider with temperature is vertical distribution along the body, particularly below the knee. These temporal and spatial factors are crucial for creating comfortable conditions conducive to concentration and thus productivity at desk work.

(3–4) Preferred conditions for the thermal environment depend on the level of body activity: for example, the bedroom is a place for relaxing and good sleep, while a kitchen space serves more active purposes. Likewise, activities are quite different for performers on the stage of an opera house or a concert hall and listeners in the audience areas. The temperature should be cooler on stage for performers and warmer in audience areas. Careful design of the thermal environment plays an important role in the resulting level of comfort and energy conservation, thereby contributing to the realization of pleasant experiences in the theater and other spaces of performing art.

(5) Utilization of night electric power to a regenerator in each house is effective in keeping temperature comfortable all day long. Use of natural solar energy and a regenerator in the daytime is particularly useful, because it is inexpensive due to a small number of consumers.

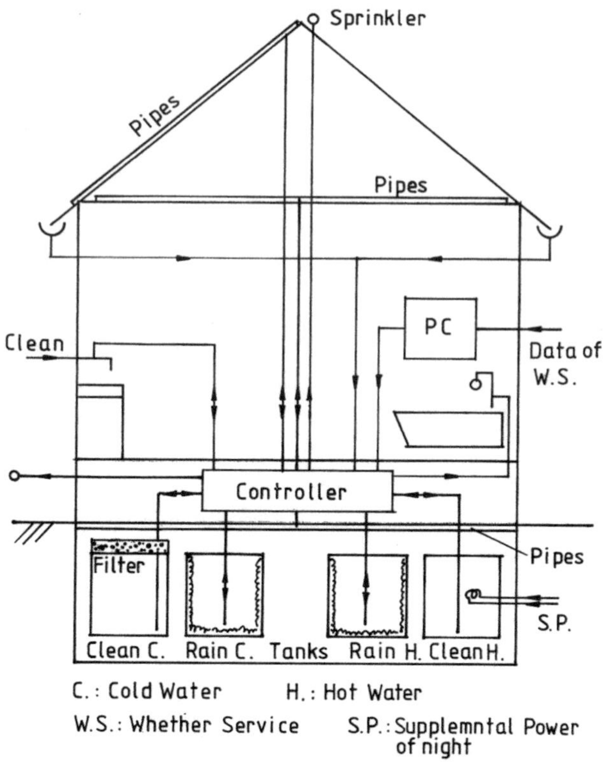

Fig. 9.4 Proposal of thermal design utilizing directly natural energies of daily and seasonal periods with tanks underground

(6) It is possible to conserve energy by utilizing thermal energy from garbage furnaces in regional areas. For example, the method could be used every weekend for warming water in swimming pools.

(7–9) It is possible to apply atmospheric phenomena including seasonal fluctuation of wind and sunshine to environmental solutions. Leaves of deciduous trees planted in the garden of a house, for example, make for a wonderful source of shadow in the summertime, and in the winter, the tree allows sunshine and energy into the house. Figure 9.4 shows an example of how it is possible to use underground tanks to store hot water collected in the summer and cold water collected in the winter. The hot and cold water is kept for about 3 months and, powered by an automatic temperature control system, can be used for baths, the roof, and the kitchen.

(10) Introducing the 1 h adjustment in summer time to utilization of solar energy and sunlight calls for adjustments in our daily life rhythm.

(11) Urban area planning is suggested to harness and control currents of air rising above the centers of town areas and channel them to produce breezes from nearby forests and rivers. This was noticed by the author during his experience of flying a glider over a village and surrounding forests near Han Munden to Kassel, Germany.

(12) Global control of climate by controlling the concentration of chemical substances in the air including CO_2 in order to sustain an inhabitable thermal environment and thereby climate. At this level, international cooperation is necessary for taking temporal design into due consideration.

9.4 Environmental Design for the Third Stage of Life

Since there are a great number of problems to be solved and most of the unknown to be discovered, each individual is hereby urged to contribute to society by his/her unique personality. In this volume, typical human creative activity is called the third stage of life that incorporates two cerebral hemispheres. The time spent in creative activity renders our mind rather insensitive to the passage of time. Such unconscious experience of duration during creative work is caused by deep concentration (Appendix A1: Ando and Ando 2010) and can be referred to as being "out of time" (Sect. 10.5). This kind of activity, in turn, serves to not only find an individual's personality and uniqueness but may well give a purpose for the individual to establish a healthy and sustainable lifestyle. Resulting creations could contribute to society and the quality of human life, helping to sustain the environment for a long time even after the passing of the individual's first (body) and second (mind) lives. A lasting peace on earth may be achieved by release of individual personality given by nature (p. 220 in Ando 1998). To support the development of the brain in babies and children, design of the environment for their third stage should be conducted with much attention, because the brain is almost developed by the beginning of elementary school. The closest surroundings at home and the neighborhood influence the development of the cerebral hemispheres to a great degree.

Figure 9.5 demonstrates design elements for the third stage of life.

(1–2) Unexpected associations and good ideas often occur in the middle of the night during sleep. In order to write down these inspirations, we need a small table and a lamp with soft light next to the bed. It is important to enable writing in a relatively dark environment to condition an easy shift back into sleep again.

Drawings hung on walls and a flame from a well-designed fireplace in the living room can stimulate the right and left hemisphere, respectively. Leaves of trees outside a window moving in gentle wind also stimulate the left hemisphere. Pleasant breezes through a well-designed window facing a green courtyard producing down currents of air may help to refresh the brain. Offering perspective stimulates the right hemisphere, which in turn plays a most important role in aiding long-term planning in individual life as well as the environment. A window with a view to and/or a viewing area for contemplating moving clouds and their scattering and vanishing in the air above the house is useful for stimulating both hemispheres: the left hemisphere is affected by the temporal movements and the right hemisphere by their form. These are all affections of nature.

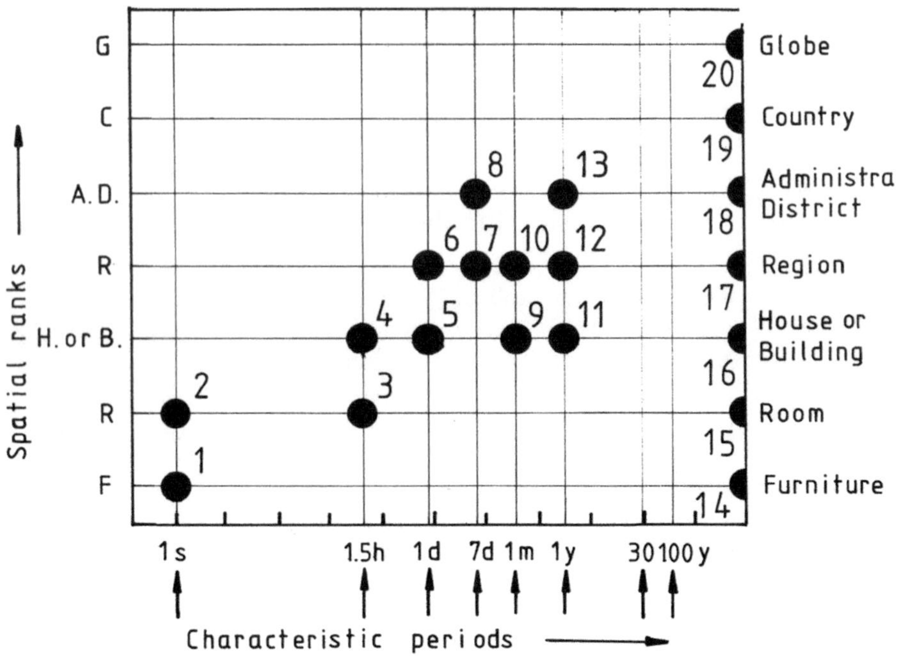

Fig. 9.5 Selected elements in the matrix with discrete periods and distinct ranks of space to be considered in the third stage of human life. Numbers *1* through *9* indicate certain temporal and spatial design projects. A tendency is observed that small spaces are designed mainly for short terms, and large spaces are designed mainly for long terms

(3–4) A creative work space (CWS) as shown in Photo 11.2 and a creative kitchen space (CKS) with designated work areas for the left and right hemispheric tasks, respectively, may well help to produce good cooking and creation. Good ideas often suddenly pop up from the mind, particularly when we switch from one task to another.

(5) Arts and books corners in a house stimulate the right and/or the left hemisphere. A garden changing with sunlight during the day and daily changes in the shape and color of the plants, for example, help to activate both hemispheres.

(6) A well-designed walkway and a riverside in a region make for ideal conditions for refreshing the body and mind, thus helping to obtain further ideas for creations.

(7–8) Weekly programs for an opera house, art museum, concert hall, church, zoo, aquarium, and botanical garden for people living in the local area. For example, if programs could be planned such that, for example, Jazz concerts are performed on Saturday evenings and classical music is performed on Sunday evenings, then people could attend the events according to the weekly schedule without extra cost for public notice of performances. In any given concert hall, musicians may compose music blending the temporal factors of the sound field

Fig. 9.6 The work "Reflection" by Marianne Jogi exemplifies temporal design merging with everyday life. The work is a small object but creation in a larger scale enables involving different settings and spaces. The picture is of the Sun reflecting off the surface of a water body, but due to certain parameters, at first glance it seems to be the Moon in deep night, reflecting the light of the Sun. Printed on glass with the light center left transparent, the object comes to life when viewed against strong natural light in the day or the night. The artifact transforms from being an object of interest into being a filter for viewing the world. The viewer simultaneously confronts herself due to the reflective surface of the glass

in the hall (typically reverberation time). The sound field then acts as a base for the musical composition. It is well known that, for example, long reverberation time in a church blends well with pipe organ music, but not with speech signals.

(9–10) A place for viewing the Moon and moonlight inside a room could be part of conscious design. An example is shown in Fig. 9.6, an art created by Marianne Jogi.

(11–12) Seasonal color change of leaves of trees may be considered at the stage of selecting the trees for a street or a park. Festival, concert, and art event venues could be adapted for every season. Soundscape design of water flow and leaves moving in the wind create sounds as a kind of music as well as "visual music." Agricultural fields are another expression of landscape, which change along with the four seasons.

(13) Seasonal programs at schools, temples, churches, and opera houses may be brought together into one weekly program.

(14–17) Accepted artistic works and scientific works might be classified as "classic" by succeeding generations. Education should pay special attention to the development of the human brain, because there are certain time windows at

Fig. 9.7 Individual possible preferred directions due to special talent with different DNA for unique creations to be determined as shown in Figure P.6

young ages (under 18 years), where the capability to accept languages, music, and mathematics is at its peak. If applied wisely, the closest environment to the human such as the furniture he/she uses or drawings he/she looks at are effective tools for the development of the brain, mind, and body.

(17–20) So-called "conservation" of nature on Earth and in the universe can be performed in terms of time.

So far, we discussed about the third stage of life in different time periods and spatial spans. Since each individual has a special talent stemming from unique DNA, it is highly recommended to investigate and try out every preferred direction at the initial stage as shown in Fig. 9.7 to save and create our environment.

The ongoing big change of global climate begs for much more attention on the dynamical and temporal design of the environment, blending human life with natural cycles, in which the third stage of life plays an important part. It is worth noting that the longest possible period of the universe is the time between one big bang and another big bang, i.e., about 10^{12} years (Appendix A.3: Steinhardt 2009).

References

Ando Y (1977) Effects of noise on duration experience. J Sound Vib 55:600–603
Ando Y (1988) Effects of daily noise on fetuses and cerebral hemisphere specialization in children. J Sound Vib 127:411–417
Ando Y (1998) Architectural acoustics, blending sound sources, sound fields, and listeners. AIP Press/Springer, New York
Ando Y (2001) Differential effects of noise and music signals on the behavior of children. J Sound Vib 241:129–140
Ando Y (2007 and 2014) Concert hall acoustics based on subjective preference theory. In: Rossing TD (ed) Springer handbook of acoustics. Springer, New York
Ando Y (2009) Concert hall acoustics and musical expressions. ARTES Publishing, Tokyo. (in Japanese)
Ando Y, Ando T (2010) Model of "unconscious" duration experience while listening to music and noise. J Temporal Des Architect Environ 10:1–6. http://www.jtdweb.org/
Ando Y, Hattori H (1977a) Effects of noise on sleep of babies. J Acoust Soc Am 62:199–204
Ando Y, Hattori H (1977b) Effects of noise on human placental lactogen (HPL) levels in maternal plasma. Br J Obstet Gynaecol 84:115–118
Ando Y, Nakane Y, Egawa J (1975) Effects of aircraft noise on the mental work of pupils. J Sound Vib 43:683–691
Bosworth T, Ando Y (2006) Design of a hillside house in Kirishima with a small office. J Temporal Des Architect Environ 6:18–22. http://www.jtdweb.org/
Fraisse P (1984) Perception and estimation of time. Annu Rev Psychol 35:1–36
Ikeda Y (1997) Designing a contemporary classic concert hall using computer graphics. In: Ando Y, Noson D (eds) Music and concert hall acoustics, conference proceedings of MCHA 1995. Academic, London, Chapter 2
Levy J, Trevarthen C (1976) Metacontrol of hemispheric function in human split-brain patient. J Exp Psychol Hum Percept Perform 2:299–312
Maki F (1997) Sound and figure: concert hall design. In: Ando Y, Noson D (eds) Music and concert hall acoustics – proceedings of MCHA95. Academic, London, Chapter 1
Mouri K, Akiyama K, Ando Y (2001) Preliminary study on recommended time duration of source signals to be analyzed, in relation to its effective duration of autocorrelation function. J Sound Vib 241:87–95
Schell LM, Ando Y (1991) Postnatal growth of children in relation to noise from Osaka International Airport. J Sound Vib 151:371–382
Soeta Y, Uetani S, Ando Y (2002a) Relationship between subjective preference and alpha wave activity in relation to temporal frequency and mean luminance of a flickering light. J Opt Soc Am A 19:289–294
Soeta Y, Uetani S, Ando Y (2002b) Propagation of repetitive alpha waves over the scalp in relation to subjective preferences for a flickering light. Int J Psychophysiol 46:41–52
Sperry RW (1974) Lateral specialization in the surgically separated hemispheres. In: Schmitt FO, Worden FC (eds) The Neurosciences: third study program. MIT Press, Cambridge, Chapter 1
Steinhardt PJ (2009) Cyclic universe theory. McGraw-Hill Yearbook of Science & Technology, New York, pp 69–71

Chapter 10
Studies on Temporal Design for Visual Fields

10.1 ACF Analysis of Camphor Leaves Moving in Natural Wind

Leaves are very common in the visual environment; thus the camphor leaf (Fig. 10.1) and a whole camphor tree (Photo 10.1) were selected for analysis. First, the images were analyzed by using the ACF to attain the physical characteristics of camphor leaves moving in the wind. Focus was on the dynamic changes in the lighting levels of the leaves' reflective surfaces and the orthogonal properties of the physical and temporal factors extracted from the ACF at different wind speeds. Secondly, to evaluate the preferred temporal properties, paired-comparison tests were conducted as discussed in the following section. Time-variant images possess a multitude of structures and regularities at many levels of complexity, including the ramifications brought about by, for example, the different dimensions of the branches and the trunk, their sequential movement, etc. Our ultimate goal is to establish a method for evaluating the temporal factors in the visual environment that affect human preferences. Such a method could be used for designing much richer environments in which the passage of time can be enjoyed (Soeta and Ando 2001).

Temporal images of windblown camphor leaves were recorded between 12:30 and 1:00 p.m. on a sunny winter day using a video camera. The images were taken on a hillside in an open space (25 × 70 m) on the Kobe University Campus. The camera was placed in front of the reflective surfaces of leaves at a distance of 1.1 m and a height of 1.0 m; the elevation angle was 30°. The wind speed was concurrently recorded. Four 10-s images with different degrees of movement caused by the wind blowing at speeds of $0.71 \pm (0.03)$, $1.15 \pm (0.14)$, $2.40 \pm (0.08)$, and $3.12 \pm (0.29)$ m/s were analyzed. Values in parentheses signify the standard deviation (SD). The images were digitized using a video capture board at a resolution of 320 × 240 pixels and a gray level of 8 bits/pixel (256 density levels). The frame rate was 30 frames per second. To find the area, in which the dynamic changes in the

Y. Ando, *Brain-Grounded Theory of Temporal and Spatial Design*,
DOI 10.1007/978-4-431-55891-0_10

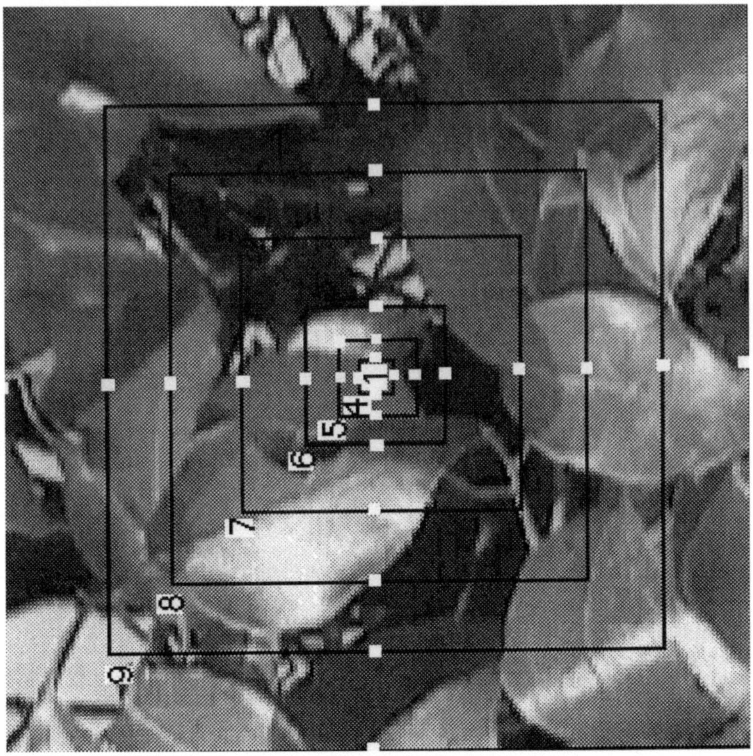

Fig. 10.1 Areas integrated corresponding to visual angles of (1) 0.01, (2) 0.02, (3) 0.05, (4) 0.1, (5) 0.2, (6) 0.4, (7) 0.8, (8) 1.2, (9) 1.6, (10) 2.4 degs. *White dots* show positions analyzed corresponding to a visual angle of 0.01

lighting level could be analyzed, the gray levels in the central area of each image were changed to (1) 1 × 1, (2) 2 × 2, (3) 4 × 4, (4) 8 × 8, (5) 16 × 16, (6) 32 × 32, (7) 66 × 66, (8) 98 × 98, (9) 130 × 130, and (10) 196 × 196 pixels, as shown in Fig. 10.1. They were analyzed by using the ACF. The actual visual angles of these areas were, respectively, (1) 0.01°, (2) 0.02°, (3) 0.05°, (4) 0.1°, (5) 0.2°, (6) 0.4°, (7) 0.8°, (8) 1.2°, (9) 1.6°, and (10) 2.4°. Results show that the ACF factors leveled off at visual angles below 0.05°. Consequently, one pixel, the minimum unit, was selected for the analysis. The gray levels at 29 points in the digitized images (white dots in Fig. 10.1) were also analyzed by using the ACF. An example of the sequential gray-level data is shown in Fig. 10.2. Compression of the gray-level data in the analysis was not considered explicitly, because no extreme high or low values of luminance were observed.

The ACF is the fundamental measure of time-sequential data. The subjective attributes may be described by one or two of four orthogonal factors that can be extracted from it (Fig. 4.13). In the study, the running ACF was analyzed for the gray level changing as a function of time. An example of a measured ACF is shown

Photo 10.1 Camphor tree

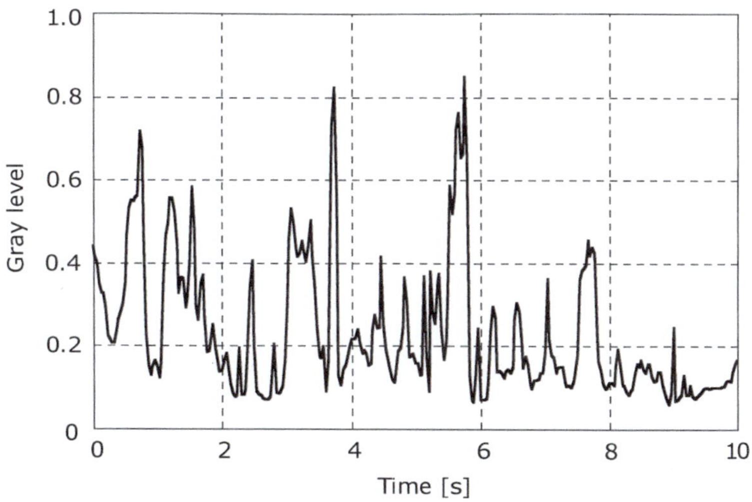

Fig. 10.2 Sequential gray-level data at a wind speed of 3.1 m/s

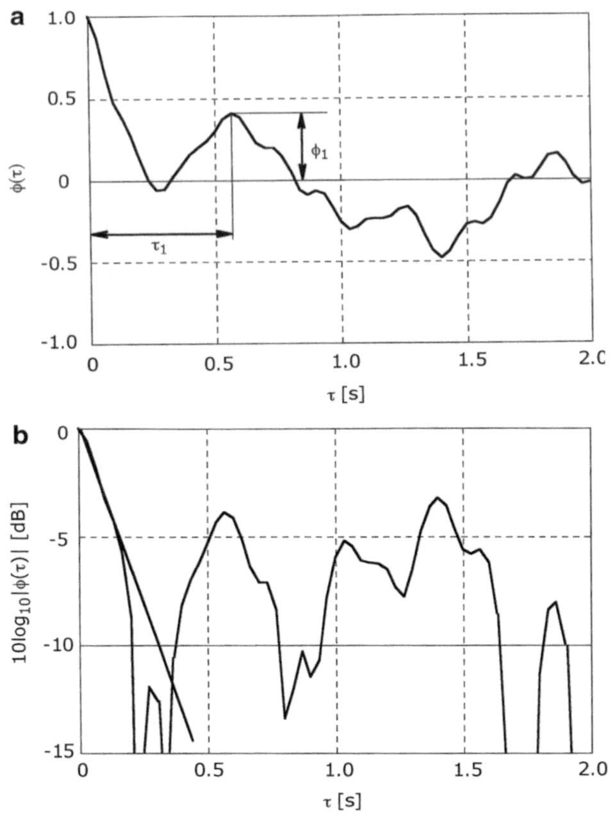

Fig. 10.3 (**a**) Example of a normalized ACF of gray level analyzed for a wind speed of 3.12 m/s and the definitions of factors φ_1 and τ_1. (**b**) An example of straight-line regression with initial part of the normalized ACF in the logarithm scale to obtain the effective duration of ACF (τ_e), which is defined as the cross point of -10 dB at given delay (the ten-percentile delay)

in Fig. 10.3a. The four factors are extracted as follows. The first is the effective duration of the normalized ACF, τ_e, defined by ten-percentile delay as demonstrated in Fig. 10.3b. The second is the energy at the origin of the delay, $\Phi(0)$. The fine structure of a normalized ACF includes peaks and dips and its associated delay times. The symbol τ_n represents the delay time, and ϕ_n represents the amplitude of the nth peak ($n = 1, 2, \ldots$). Since there was a certain degree of coherence between parameters of τ_n ($n = 1, 2, \ldots$), and ϕ_n, these parameters may be represented by the last two factors: the amplitude of the first peak, ϕ_1, and its delay time, τ_1. The values of τ_e, $\Phi(0)$, ϕ_1, and τ_1 were analyzed as a function of running time for $2T = 2.5$ s, with a 100-ms interval.

The cumulative frequencies of the four factors, τ_e, $\Phi(0)$, ϕ_1, and τ_1 at 29 positions are shown in Fig. 10.4a–d, respectively. There were no correlations between the four parameters; thus the four temporal factors are almost orthogonal in the present condition (Soeta and Ando 2001). The median (50 %) value of τ_e decreased as the

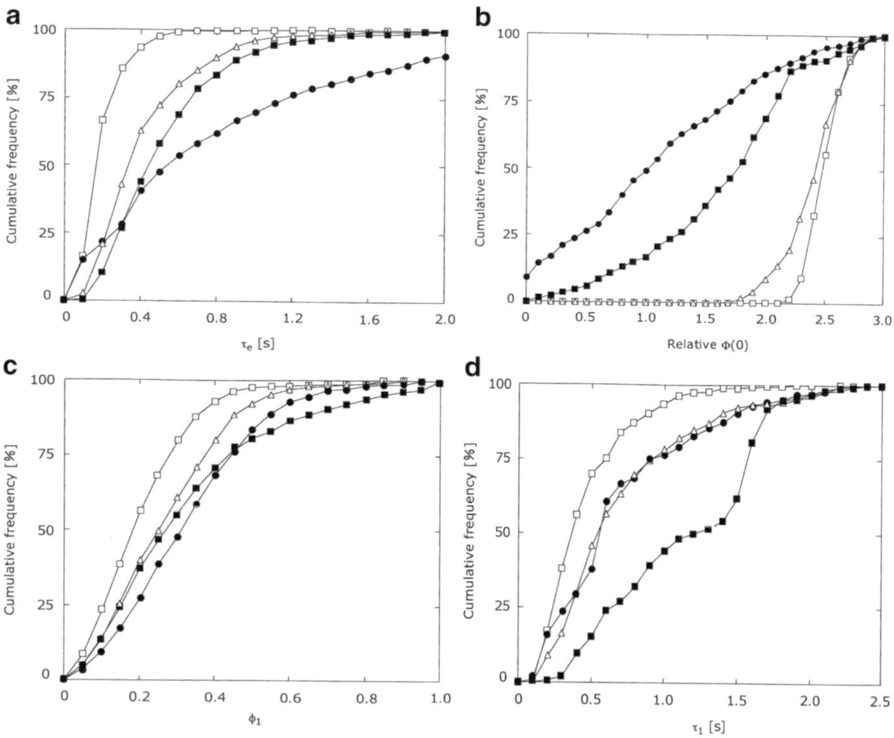

Fig. 10.4 Cumulative frequencies of (**a**) τ_e, (**b**) $\Phi(0)$, (**c**) φ_1, and (**d**) τ_1 observed at 29 positions. Wind speeds are represented by the following symbols: ■: 0.71, ●: 1.1, △: 2.4, and □: 3.12 [m/s]

wind speed increased, except at 1.15 m/s. The median value of τ_1 showed the maximum at a wind speed of 0.71 m/s and the minimum at a wind speed of 3.12 m/s. Bifurcation is observed at wind speeds of 0.71 and 1.15 m/s; thus nonlinear relationships between wind speeds and four factors were observed. It is supposed that such complicated movement of leaves influences our mind and brain at a certain level. In the next section, let us discuss subjective preferences for moving camphor leaves.

10.2 Subjective Preferences for Camphor Leaves Moving in Wind

To obtain the most preferred temporal properties of the sequential images, the paired-comparison test (PCT) on the analyzed images was conducted. Results show that the most preferred condition is described by the ACF factor representing fluctuation: $\phi_1 \sim 0.2$–0.3.

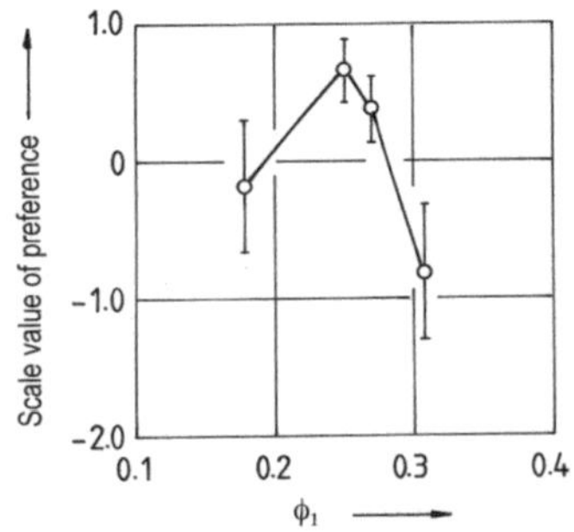

Fig. 10.5 Scale value of subjective preference in relation to the factor ϕ_1

Subjective preferences for images of camphor leaves moving in the wind have been investigated in relation to their temporal factors (Soeta and Ando 2001). The 5-s sequential images analyzed here were displayed on a CRT display placed in a dark room at a distance of 1.1 m from the subjects' eyes. Considering the focal vision, the area of the images presented was within a 1.2° visual angle. Ten subjects (22–32 years old) with normal or corrected visual acuity participated in paired-comparison testing. Images were presented at 1.0-s intervals, and the interval between comparison pairs was 4.0 s to allow time for the subjects to respond. Each subject was asked to judge which image they preferred to watch, and they responded by pressing a button. Tests were performed for all combinations of the pairs, i.e., 12 pairs; interchanging the order in each pair per session, a total of 6 sessions were conducted for each subject. Scale values based on the subjective judgment of each subject were obtained.

The global-scale value of the subjective preferences showed a maximum at wind speeds of 2.40 m/s and a minimum at wind speeds of 1.15 m/s. There was a certain degree of agreement between all the subjects, especially at wind speeds between 1.15 and 2.40 m/s. Relatively large individual differences were found at wind speeds of 0.71 and 3.12 m/s. However, no systematical relationship between wind velocity and the scale value of subjective preference was found.

After that, a study was performed to determine whether or not subjective preference is described by the temporal factors extracted from the ACF. Three 5-s images at three different wind speeds were selected, such that 0.71 ± (0.03), 2.40 ± (0.08), and 3.12 ± (0.29) m/s. The temporal factor τ_1 here might be the most relevant temporal factor extracted from the ACF of images. Median values of the four factors and a global-scale value of the subjective preferences were obtained (Soeta and Ando 2001) to determine which factors are relevant in the formation of subjective preferences. The cumulative frequencies of the temporal factor τ_1 extracted from the temporal ACF of the gray level of images at 29 positions are shown in Fig. 10.4d. For example, the median (50 %) value of τ_1 was 0.33 s at a wind speed of 3.12 m/s, 0.53 s at a wind speed of 2.40 m/s, and 1.12 s at a wind speed of 0.71 m/s, respectively.

The most remarkable results are shown in Fig. 10.5 where most of the observers preferred a certain fluctuation in the movement of the leaves, i.e., in the range of

$\phi_1 = 0.2$–0.3. However, no relationship was detected between the scale value of preference and the other factors extracted from the ACF: amplitude $\Phi(0)$ and pitch τ_1.

Our results reconfirmed that the natural time-variant scenes are not Gaussian (Dong and Atick 1995). At present stage, knowledge regarding fluctuation and regularities for time-varying images is very limited. The ACF provides information about pitch, which is represented by τ_1. It is interesting that the bifurcation of the τ_1 value represents a hierarchy of the moving leaves, i.e., the movements of the leaves and branches at a wind speed of 0.7 m/s as shown in Fig. 10.4d. When the wind was fast, the movement of the leaves had a clear pitch; when it was slow, the pitch was fuzzy. Investigation at such a level of complexity is not as simple as with laboratory experiments. However, the effective duration of the normalized ACF, τ_e, represents a kind of repetitive feature containing the signal itself. The fact that the preferred values of τ_e were found between 0.3 and 0.4 s indicates that subjective preferences can be obtained at a certain degree of repetition in the temporal image. It is notable that in sound fields with multiple reflections, the preferred delay time of the greatest reflection is deeply psychologically related to the τ_e of the source signal (Ando and Gottlob 1979). In visual fields, it is supposed that center activities in the image may correspond to the direct sound, and a coherent area within focal vision may correspond to the reflections. As a result, it is meaningful to introduce values of τ_e for subjective preference judgments in the visual environment as well.

Main conclusions obtained so far are:

1. Almost orthogonal properties of the physical factors from the ACF, i.e., τ_e, $\Phi(0)$, ϕ_1, and τ_1, were extracted for the movement of camphor leaves.
2. The preferred values may be found by the value of ϕ_1 between 0.2 and 0.3 s.
3. To some extent, the results obtained here may be related to the preferred fluctuation factor ϕ_1 as discussed in Chap. 7 and Sect. 7.1.

There are a number of different kinds of trees that produce different effects. Extreme examples are coconut trees with large leaves moving in the wind with their shadow projected on the ground or bamboo trees with fine leaves as demonstrated in Photo 10.2, which work with the wind to produce sounds also in the ultra-frequency ranges to make people relax.

10.3 Matching Acoustic Tempo with Camphor Leaves Moving in Wind

An experiment was conducted to attain knowledge on blending sound and vision. Results show that the matching period of sound pulses is deeply related to τ_1, which is extracted from the ACF of the gray level of the image of the leaves and expresses the visual equivalent of pitch.

Photo 10.2 Bamboo trees

The visual and auditory systems provide us with the majority of the information that we obtain from the environment. A number of studies have dealt with the relationship between auditory and visual perception. Sugano and Iwamiya (1999) investigated the temporal information in visual motion to determine the congruency of music and moving images. Subjects had to adjust the speed of a ball, moving in a circular or square pattern, to match changes in musical tempo.

The purpose of this study was to clarify the relationship between the temporal factors extracted from the temporal ACF of the image moving in natural wind and those of the acoustic tempo. Images of camphor leaves moving in the wind were selected as visual stimuli (Soeta et al. 2001). Pulses of sound were selected for the sake of simplicity. One of the goals was to establish a method for selecting and/or composing music that matches a lively visual environment. Such a method could be used to design richer auditory and visual environments in which the passage of time can be enjoyed.

The periods of 5-s sequential pulses used for the matching tests were 0.08, 0.16, 0.32, 0.64, and 1.28 s, with pulse widths of 63 μs (0.063 ms). Seven subjects, from 21 to 24 years old and having normal hearing and binocular vision, participated in the study. The monitor presenting the visual stimuli was placed in the front of subjects' eye position at a distance of 1.1 m to keep foveal fixation (natural binocular). The loudspeaker was placed on the monitor. The subjects' head and eye positions were unconstrained. The sound-pressure level at the center position of the subject's head was kept constant at a peak of 78 dBA. The subjects judged which sound pulses more subjectively matched up with the movement of camphor leaves. Ten pairs of combined 5-level periods constituted one series, and ten series were conducted for all seven subjects for each image of camphor leaves moving at different wind speeds.

The scale values of the matching judgment of individual subjects were obtained (Ando 1998). Figure 10.6 shows an example of the matching evaluation curve for subject E at a wind speed of 2.40 m/s, to obtain the most-matched-pulse period $[T]_m \approx 0.35$ s, which was obtained by fitting a suitable polynomial curve to the graph of the scale value.

Figure 10.7a–c shows scale values of matching for three different wind speeds as a function of normalized pulse periods for all individual subjects and the global values. It is remarkable that, similar to Eqs. (5.3) and (7.1), a matching evaluation curve may be commonly expressed by

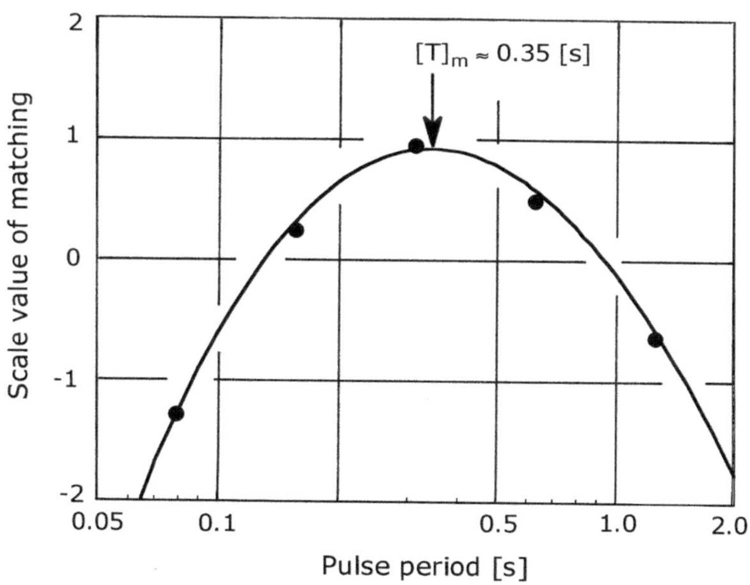

Fig. 10.6 Example of obtaining most-matched-pulse period $[T]_m$ with the movement of camphor leaves (subject E, wind speed ≈ 2.40 m/s)

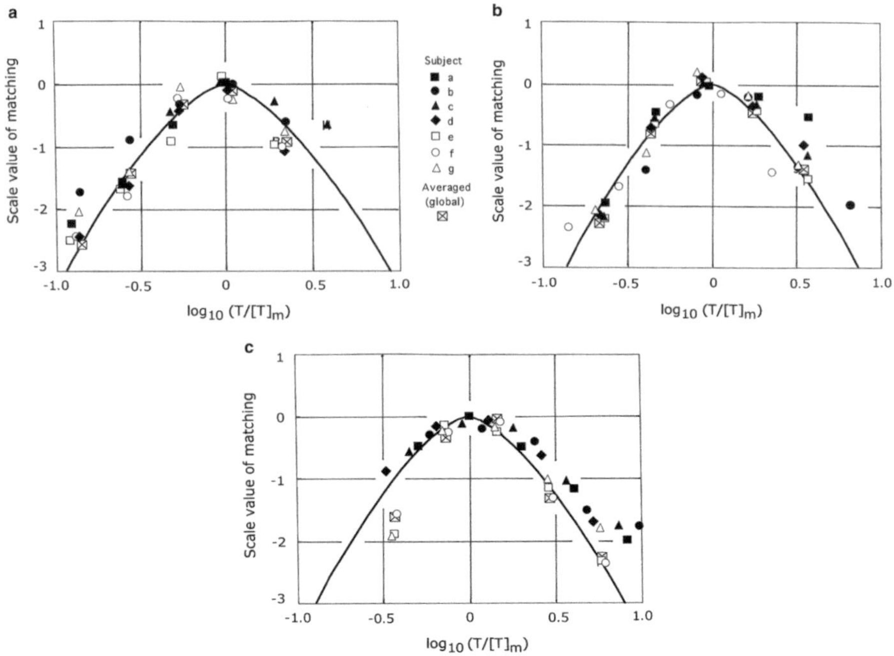

Fig. 10.7 Scale values of matching as a function of the period of pulse sounds normalized by most-matched-pulse period $[T]_m$ at (**a**) wind speed ≈ 0.71 m/s, (**b**) wind speed ≈ 2.40 m/s, and (**c**) wind speed ≈ 3.12 m/s

$$S = S_L \approx -\alpha|x|^{\beta} \tag{10.1}$$

where α and β are the weighting coefficients and $x = \log_{10}T - \log_{10}[T]_m$. After obtaining the $[T]_m$ for each subject, we identified values of α and β. The values of β estimated by using a quasi-Newton numerical method in global values were 1.18 at a wind speed of 3.12 m/s, 1.94 at a wind speed of 2.40 m/s, and 1.43 at a wind speed of 0.71 m/s. The average value of β was 1.52 (~3/2) here, thus fixed at 3/2 as usual. The solid line in Fig. 10.7 indicates the matching curve represented by Eq. (10.1) with $\beta = 3/2$. Thus, the characteristics of the matching curve can be approximately expressed by only the coefficient α, which is the sole coefficient describing the sharpness of the matching curve, indicating the strength of matching for the most-matched-pulse period $[T]_m$.

Figure 10.8 shows the $[T]_m$ as a function of the median (50 %) value of the factor τ_1 extracted from the temporal ACF of the gray level of images, which was the most relevant factor among the temporal factors. The other factors, $\Phi(0)$ and ϕ_1, were not related to the most-matched-pulse periods in this experiment. A remarkable finding is that the matched period of sound pulses is about half of τ_1 and is nearly equal to the factor τ_e. It has been discovered that the delay time of the peak of the ACF, i.e., the factor τ_1, is closely related to the perceived "pitch" of the lighting light (Sect. 3.2.1).

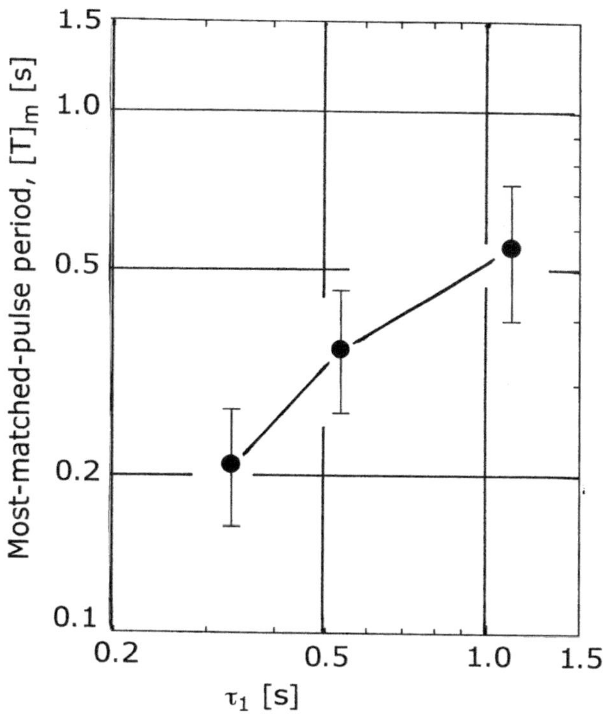

Fig. 10.8 Relationship between the most-matched-pulse period $[T]_m$ and delay time of first peak of ACF, τ_1. Ranges of the $[T]_m$ graphically obtained by all subjects at 0.1 below the maximum matching score

The weighting coefficient β in Eq. (10.1) was nearly equal to 3/2. This is consistent with any preference judgment for the sound field and the flickering light. Equation (10.1) represents the matching evaluation curve in the present study and corresponds to the preference evaluation curve of the sound field, which could mean that the theory on the subjective preference of sound fields might also be applicable to studies on the congruency of music and motion sequence.

So far, results show that the matching period of sound pulses is about half of the delay time of the first peak of the ACF (τ_1), which was analyzed from the gray level of the image.

10.4 Studies on Temporal Design for Landscape

Leaves of trees moving in the wind are a typical view of a landscape. The temporal factors that can be extracted from the ACF of an image are $\Phi(0)$, τ_1, ϕ_1, and τ_e (Sect. 4.1.1). Leaves of trees indicate complicated movements together with the branches and the trunk. For example, values of τ_1 for the movement of the trunks of coconut trees are several seconds long; on the contrary, values of τ_1 for the movement of their leaves are in the order of the millisecond, and values of τ_1 for the movement of the branches fall in between these two extremes. Therefore, a

Photo 10.3 Combination of leaves of coconut trees and those of other different kinds of trees

combination of three different movements may image us a kind of "visual music" consisting of higher, middle, and low pitches just like in a melody. A combination of leaves of coconut trees and those of other, different kinds of trees with relatively shorter values of τ_1 (high pitch due to a small size of leaves) may create something like ensemble music (Photo 10.3). Most leaves moving in the wind have mainly three pitches (τ_1) with different fluctuations (ϕ_1) due to the physical properties of (1) the leaves, (2) the tree branches, and (3) the trunks. These characteristics may be considered in the temporal design of a landscape, where natural wind is acting as a "performer," playing the "tree instruments" with different speeds of wind. Similarly, we can enjoy elements like the water flow of a river with the water surface fluctuating over the stones resting at the bottom of the riverbed, or a fire flame, or a breeze.

Breezes can be realized by a combination of a colony and surrounding forests, where solar energy on the roofs of houses creates a current that rises higher over the colony, making space for a cooler current, i.e., a breeze from a nearby forest.

10.5 Drawing with Spatial "Vibrato" as an Artistic Expression

A suggestion is offered here for further works to clarify that by the artistic expression of Paul Cézanne (1839–1906), the first remarkable attempt was made at harmonic color distribution throughout the canvas. Color modulation as well as sequential form in a drawing might be created by the spatial factors extracted from the spatial ACF of an image.

Artistic expressions may be found in Photo 10.4, which was personally provided by Werner Lauterborn in 2003 and photographed by Gisa Kirschmann Schroeder at the Third Physics Institute, University of Göttingen. This painting was composed and interpreted by Lauterborn based on an inspiration from cave paintings dating back approximately 18,500 years that were discovered in the cave "Cosquer" near Marseille, France, which is accessible only via an underwater entrance by diving (Clottes and Courtin 1995).

It is interesting to point out the fact that color modulation, expressed by Lauterborn as a form of "vibrato," is present in the background, which reminisces a kind of cave wall. The horsehair sequence in "vibrato" in the spatial frequency may be analyzed by the ACF of images with the gray level. There is a certain degree of similarity as mentioned in Sects. 7.1 and 7.3, which is discussed both in the time-domain analysis of the flickering light and in the spatial domain relating to subjective preference, respectively. The horsehair sequence is periodic, neither

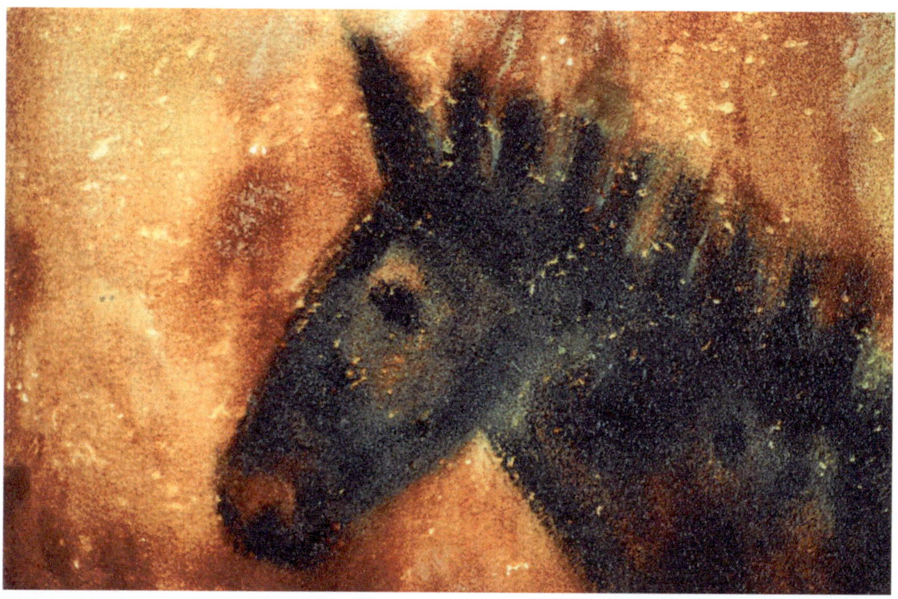

Photo 10.4 An artistic expressions of the horsehair sequence with a "vibrato" in the spatial frequency by Werner Lauterborn (2003)

perfect nor just random, and there is a certain degree of fluctuation like *vibrato*. In fact, the factor ϕ_1 extracted from the ACF is around 0.3, which is a range of the most preferred condition, depending on the individual. Regarding such color modulation including the effect of gray level, further attempts could be made to investigate the physical spatial factors with multiple-dimensional correlation analysis.

Lauterborn said that Photo 10.4 was painted in the years between 2001 and 2003. As a matter of fact, ideas of such a color modulation and a vibrato in the periodic structure came out without his conscious participation in the process. He could not identify how he painted it, perhaps because often, ideas of beauty seem to come from "out of time" or "beyond any specified time."

Therefore, this process may relate to the "third stage of life" (Fig. P.2), which indicates that it may survive longer in human culture, i.e., even after the life that governed the first and second stages of life has passed. Remarkably, there is a so-called time tunnel exit in the third stage of life.

The duration experience in the second stage of life (mind) depends on stimuli from the outside (Appendix A1) and aging (Appendix A2). For example, Appendix

Table 10.1 Effective environments for the 1st through 3rd stages of life, governed by individual preferences for each movement

Movement	1st stage of life	2nd stage of life	3rd stage of life
Before going to school			
Nursing	**X**	**X**	X
Sleeping	**X**	**X**	**X**
Native language	X	**X** (left)	**X**
Self made toy	**X**	**X**	**X**
Clay work and painting	**X**	**X** (right)	**X**
Children's meeting	**X**	**X**	**X**
TV and computer games	X	X	X
Natural playground	**X**	**X**	**X**
Music	X	**X** (left)	**X**
Reference and reading	X	**X** (left)	**X**
Nursery school/kindergarten	X	X	X
Foreign languages	X	**X** (left)	**X**
Elementary mathematics	X	X	**X**
Schools and after			
Primary school	X	X	X
High school	X	X	X
Partner	**X**	**X**	**X**
Cooking	**X**	**X**	**X**
Marriage	**X**	**X**	**X**
Sexual reproduction	**X**	**X**	**X**
Walking	**X**	**X**	**X**

Note that the third stage of life based on DNA may be developed especially during the period from the time of birth to before going to school

A1 describes a condition in a living area where noise level was about 75 dBA, 80 % of the people felt that time was shorter by 0.4 times on average. If we expand this model (Fig. A2.1) to cover 90 years of lifetime, then our feelings might only reflect an experience of 36 years.

The lifetime of the first stage (body) is related to telomeres in the chromosomes (ex. Witzany 2008) and to long-term stresses.

References

Ando Y (1998) Architectural acoustics, blending sound sources, sound fields, and listeners. AIP Press/Springer, New York

Ando D, Gottlob D (1979) Effects of early multiple reflection on subjective preference judgments on music sound fields. J Acoust Soc Am 65:524–527

Clottes J, Courtin J (1995) Grotte Cosquer bei Marseille, Eine im Meer versunkene Bilderhoehle, translated into German by Uebersetzt von Kathrim Wuest, Jan Thorbecke Verlag

Dong DW, Atick JJ (1995) Statistics of natural time-varying images. Network 5:517–548

Soeta Y, Ando Y (2001) A pilot study of autocorrelation analysis and subjective preferences of images of camphor leaves moving in the wind. J Temporal Des Architect Environ 1:6–11. http://www.jtdweb.org/

Soeta Y, Uchida Y, Ando Y (2001) Matching a tonal tempo with camphor leaves moving in the wind. J Temporal Des Architect Environ 1:21–26. http://www.jtdweb.org/

Sugano Y, Iwamiya S (1999) Effects of synchronization between musical rhythm and visual motion on the congruency of music and motion picture (in Japanese). J Music Percept Cogn 5:1–10

Werner Lauterborn (2003) personal communication

Witzany G (2008) The viral origin of telomeres and telomerases and their important role in eukaryogenesis and genome maintenance. Biosemiotics 1:191–206

Chapter 11
Studies on Temporal Design of an Individual House

11.1 House with Creative Work Space (CWS)

All healthy creations based on the unique personality of the individual may contribute to human life for a long time as culture. It is highly recommended, therefore, that the environment be designed for the following three stages of human life explicitly:

1. The first life (body)
2. The second life (mind)
3. The third life (personality-based creation)

Designs for the body and mind are well known, and these stages are what humans have in common with animals. However, the third stage of creation is only characteristic to human life. A well-designed environment would be a meeting place for art and science associated with both hemispheres, which in turn, might help to discover that the individual personality is, in fact, a minimum unit of human society. Any person can find some time in the day to think about what their unique personality is like and what the most interesting thing a person can imagine him- or herself doing in life even after retirement would be. For many people, retirement means loss of hopes and dreams to do anything other than walking, reading, taking journeys, and doing hobbies. However, if a person is aware of the reality of the abovementioned third stage of life, which is very human, close to our nature, and worth aspiring to, then it might not only keep the body and mind of the individual healthy but also contribute to maintaining the environment and promotion of peace.

For the third stage of life, particularly, an example of a creative workspace (CWS) that activates both cerebral hemispheres by incorporating into the design two different panels for the left and right hemispheric activities has been designed as shown in Photo 11.1a. There is another CWS for the wife or partner as shown in Photo 11.1b. Namely, there are "three different panels" specialized for the left and

Y. Ando, *Brain-Grounded Theory of Temporal and Spatial Design*,
DOI 10.1007/978-4-431-55891-0_11

Photo 11.1 (**a**) Study with a creative work space (CWS). The plate on the left-hand side was used for the right hemispheric task such as drawing figures and preparing photographs, the center plate was used for writing manuscripts, and the right-hand side was used for the internet system. (**b**) Smaller type of CWS used by the housewife

right hemispheric tasks and for the integration and reception of information through an information-communication system between individuals and facilities such as libraries and museums. The left hemisphere is specialized for tasks as temporal processes including writing, reading, speech hearing, calculation, and logical consideration. The right hemisphere is specialized for tasks as spatial processes including pattern recognition, space forming, drawing, painting, clay work, and making scale models. Living in this house for about 5 years from March 2004–September 2009, the author, for example, could prepare the manuscripts of several volumes and separate chapters (Ando 2007, 2009a, b, c, 2011).

It is promising that multidimensional ideas may be obtained by introducing such workspaces with changing positions for different tasks. It is well known that bright and spontaneous ideas often "pop up" between changing tasks. Changing between working panels helps to create order and peace in the mind which is a prerequisite for drawing up spontaneous creative ideas. This approach is quite different from the usual one-dimensional working environment, which is likely to create only "one-dimensional ideas." Many of us have probably experienced such undesigned work environments where ideas tend to be similar and there is no "feeling" of creativity being nurtured or cultured.

It is particularly useful for the developement of a child's brain to foster the different kinds of work processes in a CWS. For example, right hemispheric tasks including clay works require multidimensional thinking. Performance may not be well accomplished on the usual single-type table or L-type table. An internet system between the home and office is likely to play a role in achieving maximum effects with minimum effort, thus saving time and energy, since there would be no need to attend the working and/or studying place every day.

11.2 Design Study of a Hillside House

Models that involve values from seemingly different ends of the spectrum, such as saving energy and simultaneously realizing delightful visual environments that evoke subjective preference and the corresponding physiological changes, fully incorporate natural light and natural energy in the design, considering the cycle of day, month, year, and generation (about 30 years) (Bosworth and Ando 2006). It is noteworthy that engagement in any natural activity (or cycle) may greatly help the development of individual personality and, therefore, creation.

The plan of the house with two axes is shown in Fig. 11.1a. In order to demonstrate temporal design that cultivates the three stages of life, a hillside house was built in Kirishima, Kyushu, Japan, about 700 m above the sea level as shown in Photo 11.2a, b (Bosworth and Ando 2006). Figure 11.1a, b, respectively, show the plan and elevations/sections. For the first stage of life (body), for instance, the bedroom is designed with a window partitioned into three smaller ones, creating a sort of cave-like space with a diminished amount of natural light reaching the room, helping the body to relax and to prepare for and enjoy sleep. For the second stage of life (mind) and particularly for the periods of 1 day and four seasons, windows in the living room, kitchen, and bathroom are carefully designed to have enough natural light from the outside, and from the inside, they enable viewing trees and the large-scale natural landscape including the Sakura-jima with an active volcano and the bay (Photo 11.3). Skylights luminate the porch and the table there. The veranda is a joyful space for tea and food in the morning and afternoon.

The dominant east-west axis is the backbone of the house and is placed parallel to the contours of the steeply sloping site. Situated along this axis are the human activities associated with the time frames of the body and the mind. This axis stretches from the dark quiet of sleep in the bedroom in the west end (Photo 11.4) through the areas for washing (Photo 11.5), cooking (Photo 11.6), eating, and socializing (Photo 11.7). Between washing and cooking, there is an aisle with scattered skylights (Photo 11.8) which then proceeds and becomes the brightness of the porch (Photo 11.9) with natural light from three sides and finally continues through space, reaching into the infinity of the rising sun in the east (Photo 11.10). The secondary north-south axis begins at the entrance porch on the north flank of the house (Photo 11.1a), progresses through the house crossing the main axis under

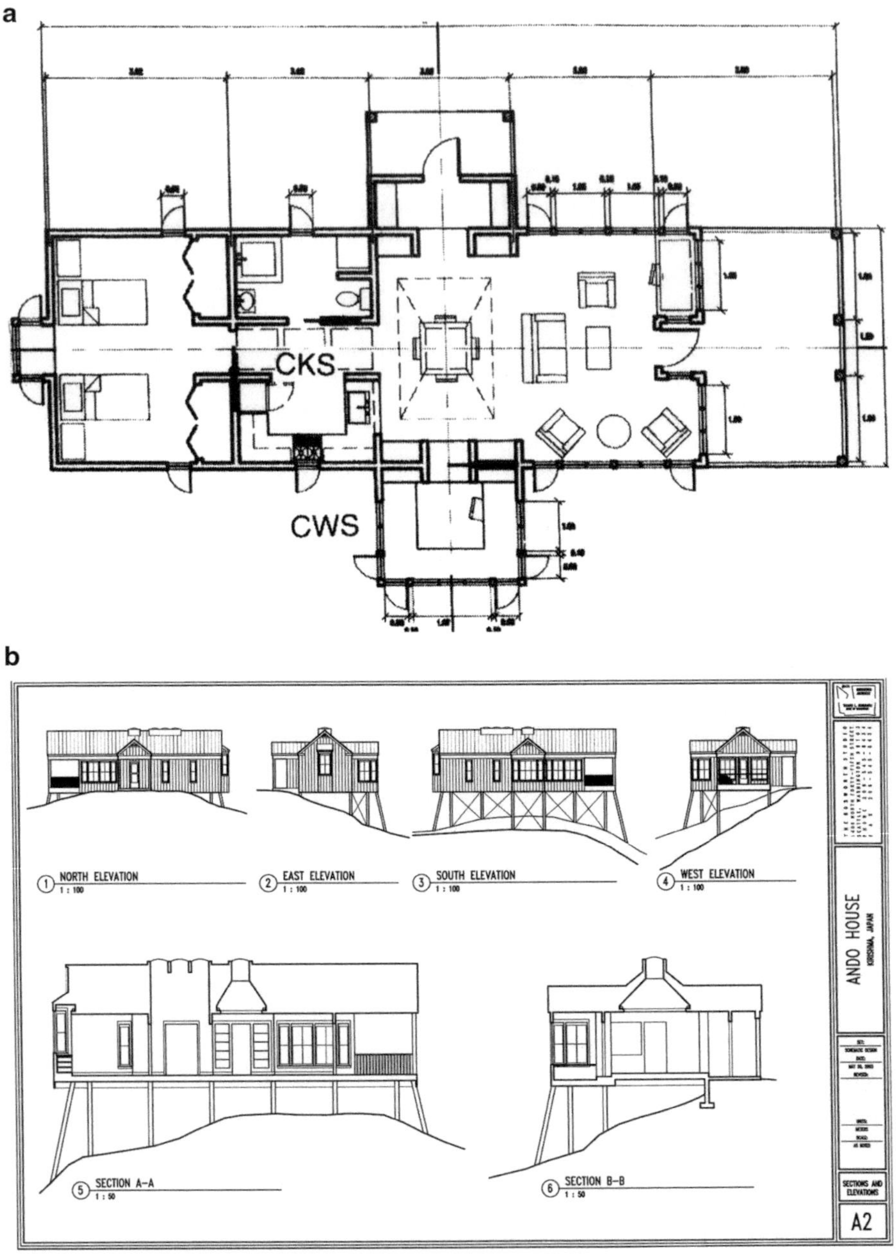

Fig. 11.1 (**a**) The floor plan of the hillside house including the CWS. (**b**) Elevations and sections of the house (Bosworth and Ando 2006)

Photo 11.2 (**a**) Exterior view of the hillside house in Kirishima. (**b**) Extrerior view of the rear of the house

Photo 11.3 A view from the porch below one's eyes: the Kirishima International Music Hall, quiet Sakurajima Volcano, and Kaimon-Dake about 150 km away

the central skylight, and terminates in the study attached to the south flank of the house (Photo 11.1b). This room is suspended in space over the sloping hillside that drops away below it. There are windows on three sides, and it is organized into three work areas (each with its own window and unique view) according to the design principles developed by Tom Bosworth, which facilitate creation by a unique personality. The body of the house is composed of traditional forms which envelop the two organizing axes, with support on legs of varying length, which stand firmly on the sloping site. Because of its upright rather than seated posture relative to the ground, the house is imbued with life which reflects as well as contains the vital human activities within. Natural light, by its daily and seasonal fluctuation and by its sensitivity to changing weather conditions, reveals and defines the exterior form and interior spaces of the house. The variations of the quantity and quality of this light and the accompanying shadows are constant reminders to the inhabitants of their own animation as participants in the flow of time. It is hoped that the survey presented here, taking into account both temporal factors associated with the left hemisphere and spatial factors associated with the right hemisphere in design of architecture and the environment, can suggest a suitable line for further works of each individual.

Photo 11.4 West window of the sleeping room

Photo 11.5 (**a**) Small window in the bathroom and toilet

Photo 11.6 View of kitchen, namely, creative kitchen space (CKS)

The creative work space (CWS) is composed of side panels and fully activates both hemispheres (Ando 2004). The porch is a space that allows the inhabitant to look at leaves and branches of trees moving with the wind (Photo 11.9) and notice birds (Photo 11.11), while after a short break, the gaze can switch to the large-scale natural landscape including the Sakurajima with an active volcano and the bay (Photos 11.3 and 11.10). An occasional gentle breeze on the veranda is pleasant for anyone enjoying a cup of tea or a meal in the morning or afternoon; therefore, in the summer, this natural phenomenon acts as a typical cooling system without using any extra energy. A breeze can also be created by a fan in the living space by applying the preferred temporal factor $[\phi_1]_p = 0.3$–0.5, which can be extracted from the ACF of air movement with fluctuation.

In a way similar to the CWS, a creative kitchen space (CKS) can be realized. Daily and weekly menus could be developed by the left hemisphere mainly, while cooking is performed by the right hemisphere.

In the design of a regular house, the following thermal environment design considerations have been realized:

1. Between the exterior wood wall and interior diatomite wall, a layer filled with fireproofed and chipped old newspapers, covered with a special vinyl plastic bag, was placed.
2. All windows have three-layered panes of glass for energy efficiency.

Photo 11.7 Eating and socializing

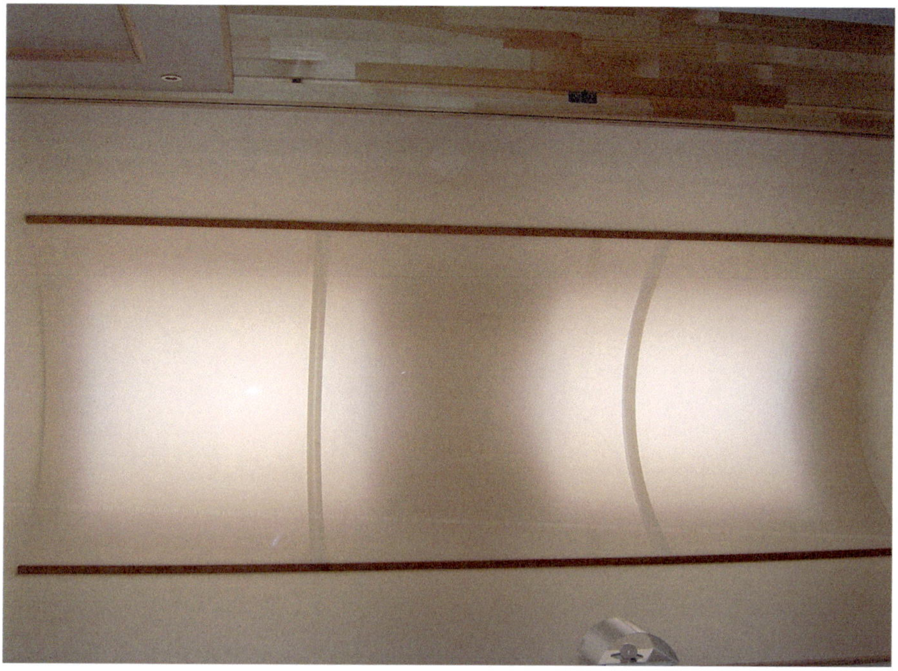

Photo 11.8 Aisle with scattered skylights

3. Under the roof, a layer of glass fiber was added.
4. Under the floor, about a 30-cm-thick thermal isolation layer made of glass fiber and Styrofoam was added, which was later increased to a 40-cm-thick layer to improve thermal isolation.
5. Two functional hot-water systems were built: (1) a natural hot-spring supply system to use for bath and partially under the floor for heating and (2) an electrical hot-water supply system for bath and kitchen that utilizes night power only. Such preferred thermal environments are enjoyable for the first and second stages of life where groundwork is laid for the third stage of life or creative works.

In this house, there are no rooms planned for children or guests, but it is easy to add extra rooms at the ends of the two axes according to need.

Photo 11.9 View of east porch

Photo 11.10 East porch with a table

Photo 11.11 Feeder near the bamboo fence for communication with birds. Every morning they would be waiting for food, and sometimes they would peek inside the house, where we were

References

Ando Y (2004) On the temporal design of environments. J Tempor Des Architect Environ 4:2–14. http://www.jtdweb.org/journal/

Ando Y (2007 and 2014) Concert hall acoustics based on subjective preference theory. In: Rossing TD (ed) Springer handbook of acoustics. Springer, New York

Ando Y (2009a) Auditory and visual sensations. Springer, New York

Ando Y (2009b) Theory of temporal and spatial environmental design. In: McGraw-Hill yearbook of science & technology. McGraw-Hill, New York, pp 384–389

Ando Y (2009c) Concert hall acoustics and musical expressions. ARTES Publishing, Tokyo (In Japanese)

Ando Y (2011) Brain oriented acoustics. Itto-Sha, Tokyo (In Japanese)

Bosworth T, Ando Y (2006) Design of a hillside house in Kirishima with a small office. J Tempor Des Architect Environ 6:18–22. http://www.jtdweb.org/

Chapter 12
Development of the Third-Stage (Creative) Life

12.1 Annual Drama-Performance in a Village

In a village, voluntary young men (about 16–20 years of age) started to perform classical drama (Kabuki) during autumn harvest. It is called Noson Kabuki in the east of Mimasaka and originates from the middle of Edo Era about 300 years ago. Beginning the preceding winter, about 7 or 8 months prior to the planned performance, one of the carpenters of the village, whose name was Takeo Nishikawa, took up a hobby and began to teach the lines to the performers without taking charge of the production of the event. For such performances to be played out, a small stage house was built more than 100 years ago. When the author was in his elementary school years, 2-day performances were held each year. Before the performing days, the audience floor area on the farm ground as well as the side and rear walls would all be temporarily covered with straw mats, and an elevated passageway would run from the left side of the stage to the rear part, and from there to a wood cabin located on the right side just in front of the stage that is meant for the performance of music on a shamisen (Japanese music instrument with three strings). Therefore, the theater is a half-open space, offering the possibility to enjoy the Moon, stars, and sometimes even shooting stars in the night sky.

On a particular evening, the drama began in the evening with a traditional prelude named Sanbaso. According to the skill of the performers, the audience had the chance to contribute small amounts of money at the reception desk. A notice was provided on the wall near the reception with names of the performers, amount of money, and the names of the contributors specified on it. The author's elder brother Toshio Ando played the role of an Oyama (a female role played by a male actor) and was selected the top actor who received many contributions. Many people brought along picnic boxes with homemade Saba-zushi, Maki-zushi, and Inari-zushi, but not Nigiri-zushi due to the fact that it is hard to obtain fresh fish in a village located on a mountain side.

Y. Ando, *Brain-Grounded Theory of Temporal and Spatial Design*,
DOI 10.1007/978-4-431-55891-0_12

Pupils of elementary and junior high school watched and listened to the performances that they themselves would be giving in the future. Often on their way to and back from school, they enjoyed pronouncing highlighted phrases from the drama. They recognized that "All the world's a stage," as Shakespeare said.

Nowadays, due to the development of TV broadcasting and the fact that young people have left the village and moved to town areas, Kabuki in that particular village is no longer performed. However, in four villages in the East Mimasaka area, performances are still going on.

12.2 Weekly Meeting of Children in a Village

Once a week in the evening, in a room of an individual house, children between the ages of 10 and 15 organized a meeting for children without parental guidance. All presenters were elementary and junior high school children. Parents participated as galleries who did not give any advice or suggestions but simply enjoyed the performances of the children.

The children prepared joyful slides written on glass film in black ink, cut at a width of about 6 cm to be displayed on their self-made slide projector with an incandescent electric bulb (100 W) as shown in Fig. 12.1. An example of a self-made glass slide is shown in Fig. 12.2. After presentation, figures on the glass could be easily erased with water, and new figures could be drawn once the glass was dry. Without any further costs, children could enjoy creations such a MANGA on the glass. At the beginning of the meeting, figures were magnified and displayed on a FUSUMA (an architectural room partition made of layered papers about 2 cm thick with pictures on the surface). Sometimes, the fusuma projected notifications about

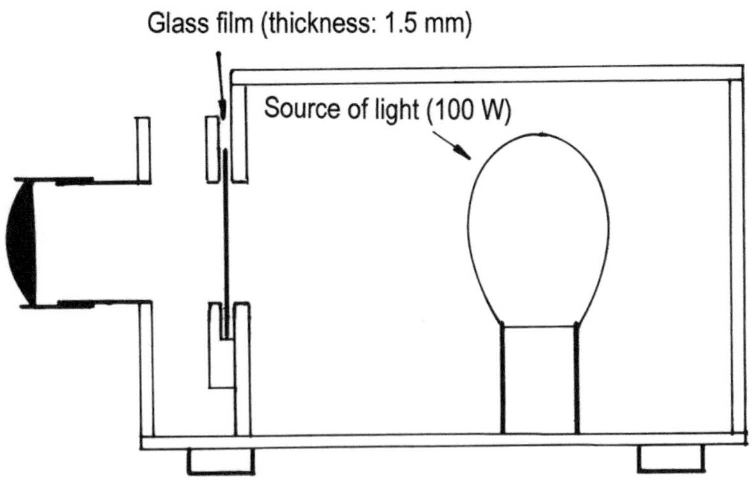

Fig. 12.1 Projector made by children in ca. 1953

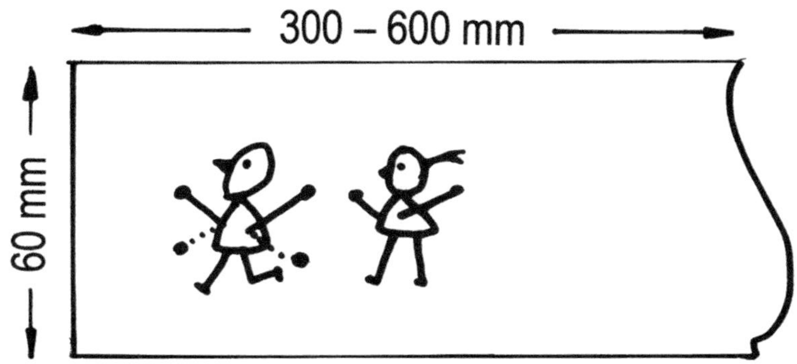

Fig. 12.2 Art works on the glass of 1.0 mm to be displayed on walls (length of about 40 cm and width of 6 cm)

the schedules of the next children's meetings. About 10 children participated in the meetings, and some parents with children under school age and neighbors took part as watchers. Many enjoyable programs were performed by school children, such as shadow plays with fingers utilizing the light of the projector, rendering foxes, dogs, and mushrooms, for example. We also enjoyed pantomimes, riddles, puzzles, and singing songs.

12.3 An Example of Prompt Creative Talk

When the children were getting tired, someone asked them to tell a prompt and creative story. It went like this: "One day when the sky was bright and clear, we started to walk towards the west along a narrow pathway. On the left-hand side the small Hino-kami Shinto Shrine appeared (Drawing 12.1, "Sprite of a tree"). We continued to walk and after a short while, a brook with a beautiful stream and water grass in the bottom could be seen (Drawing 12.2, "Stream"). We continued wandering across a small bridge and saw a water wheel (Drawing 12.3) on our left that gave life to the village. Finally, we reached a large river, but the bridge had been lost to water brought about by heavy rain some weeks earlier. All that was left was a narrow panel floating above water, and when we walked on the vibrant panel bridge, a gentle breeze passed by and we saw clear water flowing under it..."

12.4 Play Area for Children

Why are small streams in a garden so fascinating for children? A small stream is an interactive "toy," changing its course and speed as the child performs small-scale engineering projects on it, blending children and water in the flow of time. Children

Drawing 12.1 "Sprite of tree" (by the author, 2012)

Drawing 12.2 Brooklet (by the author, 2012)

Drawing 12.3 Mill (by the author, 2012)

like to play with water, mud, and trees (Photo 12.1), partially because our life originated in the soil, the water, and the forest. It is easy to understand why the play space designs of recent years hold so much more interest than the "playsets." Wood structures, flexible bridges, and fanciful designs allow for much freer play of imagination than the hard surfaces and discrete steel frames of playground toys. The "toy" itself changes due to the child's actions and indeed continues to change even after the child stands back to admire his or her work.

Other environmental elements in addition to water include different types of trees, soil, and fire as well as celestial bodies such as the Sun, the stars, and the Moon. Looking at the Moon may help to deeply admire, for example, the piano sonata "Moonlight," no. 14 in C sharp minor op. 27, no.2. composed by Ludwig van Beethoven (1770–1827).

For children to recognize themselves from among their friends, sound reflectors like walls of buildings close to the play area that reflect their voice can produce further delightful sounds, making them feel and act as if they are spending time on a so-called stage. Children are unconsciously prone to liking sounds reflected back to them from the surroundings to any play area because it makes them feel like they are actually affecting their surroundings with their voice, hand claps, or other sounds they are producing. The most usual sounds they produce are voice and speech related during play with friends on the "stage."

Photo 12.1 Children playing with a tree

We now discuss design of the environment of a play area as a "stage" for the development of body, mind, and creativity of children. As mentioned above, wooden structures, flexible bridges, and creative designs allow for much freer play of imagination and creation than hard, fixed surfaces. Thus, environments that are well designed in terms of time are most likely to play an important role in the development of children.

It is worth noticing that there are certain "temporal windows" throughout the development of the brain, which signify times when a child is most absorbent for knowledge related to different languages, mathematics, as well as music performance and its composition.

Each individual has his/her own DNA and after birth, the preferred living environment of each individual is different. Therefore, each individual develops a different personality and a different viewpoint for looking at the world. In other words, every individual is born with a special appointment. Rather than seeing this as something that needs to be levelled, let us look at it as a source for creation. Healthy creations conceived not for the purpose of war but for lasting peace through the contentment of each individual will remain in the human society for a long time as the third life even after the end of the first and second lives.

Chapter 13
Creativity-Inducing Environments for Building Culture

13.1 Individual Creations

13.1.1 Creative Work Space (CWS)

Eight systems of the creative work space (CWS) were introduced to Ando Laboratory, Kobe University, in 2002. A CWS consists of three different panels specialized for the left and right hemispheric tasks (Ando 2004, 2009a, b) and for the integration of knowledge by utilization of an information-communication system. Left hemispheric tasks are the temporal processes including writing, reading, speech hearing, calculation, and logical considerations (Sperry 1974). Right hemispheric tasks are the spatial processes including pattern recognition, space forming, drawings, paintings, clay work, and making of scale models (Fig. 13.1). It is promising that multiple-dimensional ideas can be created by such multilayered work spaces. This concept is quite different from the usual one-dimensional working space, which yields only "one-dimensional ideas." It is quite natural that people subjected to similar conditions create similar ideas in such one-dimensional spaces. As listed in Table 13.1, eight users of the CWS system were reported having work experiences better by 2.5–15 times (mean value: 5 times) than behind one-dimensional desks, which they have used, and work efficiency increased by 2–15 times (mean value: 3 times). In total, all users reported increased efficiency by at least 2 times ($p < 0.01$). Using the walls around the three panels to display verbal and nonverbal materials previously created by a user may induce further creations. Photo 13.1a is an example of the CWS activating two cerebral hemispheres in the laboratory, and Photo 13.1b demonstrates a meeting of PhD students for exchanging knowledge, discussing mutual problems, and enjoying a cup of tea after work and concentration at individual CWSs.

Y. Ando, *Brain-Grounded Theory of Temporal and Spatial Design*,
DOI 10.1007/978-4-431-55891-0_13

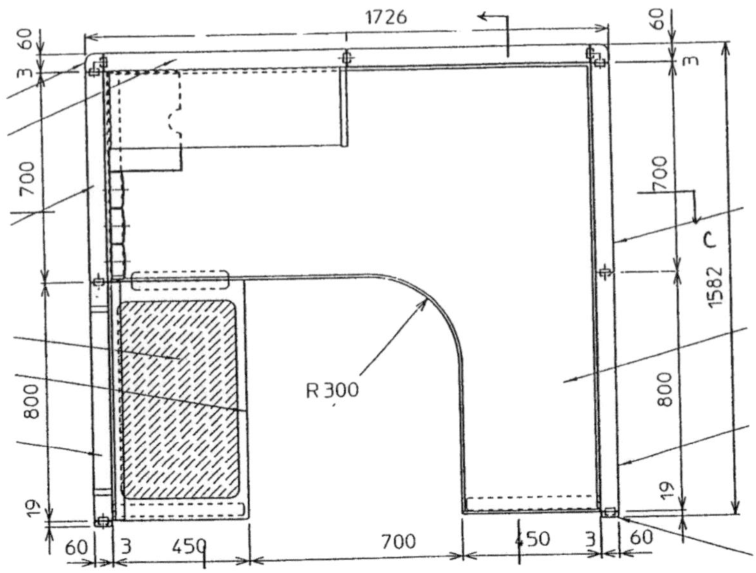

Fig. 13.1 CWS introduced to Ando Laboratory, Graduate School of Kobe University

Table 13.1 Overall work efficiency and evaluations for the CWS with reference to the usual office desk judged by eight users

User	Overall evaluation (time)	Work efficiency (time)
RS	5.0	3.0
YS	2.5	2.5
YO	10.0	5.0
TH	5.0	3.0
KK	15.0	15.0
KF	5.0	3.0
YA	3.0	3.0
SS	10.0	2.0
Average	6.9	4.6
Mean (50 %)	5.0	3.0

13.1.2 Integrating Knowledge

A vivid example to illustrate the importance of temporal design is about a new town (e.g., Senri-Newtown near Osaka) that was built as a satellite town mainly for young couples and families. Within about 30 years, the town grew old as no more young people would move there. The kindergarten was no longer needed and only senior houses were necessary. One reason why this occurred can be explained in

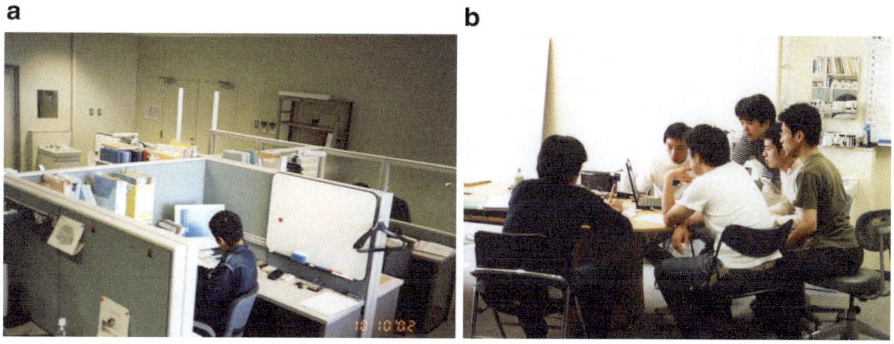

Photo 13.1 (**a**) CWS introduced Ando Lab., Graduate School of Science and Technology. (**b**) Table for discussion placed near a set of CWS

terms of temporal design, or lack thereof: there were no facilities for integrating knowledge, such as a museum, concert hall, or library, which constitute spaces and venues for refreshing the brain and inducing creations. It is natural for the human being to be driven toward living in an environment that evokes creativity and contentment because it is the natural, life-maintaining state of the human physiology. We have an innate preference for living in an environment that supports the third stage of life. Therefore, before construction of a town, urban planning should be carefully thought out in terms of temporal design for the periods of a generation of about 30 years, and a lifetime of roughly 90–100 years. Environment is a reflection of our psychology and physiology, and vice versa; therefore, people have a need for finding this world worth living in. This could be the third stage of life as defined in Preface, i.e., personality-based creations with affection toward every creature in the world.

This process, being oriented toward maintaining life, may keep the body and mind healthy and active until the end of life.

13.2 Concert Hall

For the purpose of enhancing individual creativity of the performers and audiences, the sound field at each seating position in a concert hall was designed according to preferences for both groups. The preferred conditions of temporal and spatial factors of the sound field may be designed (Chap. 6) *according to the theory of subjective preference* (Sect. 5.1, Ando 1983, 1985, 1998).

Scale values of subjective preference for all orthogonal factors are integrated by Eq. (5.2). Scale values for each seating position are calculated based on the architectural scheme under design, so that the total scale value is obtained, and consequently scale values for an alternating scheme are obtained. This process is repeated until satisfactory results are attained. Music performers are advised to blend the selected program with the temporal factors of the sound field of a given acoustic space to achieve optimal results.

To obtain an excellent sound field, the genetic algorithm (GA) operation was conducted on a shape modified from the shoebox (Ando 2007, 2009a, b). The initial form of the hall was a shoebox: 14 m wide, the stage was 9 m deep, the room was 27 m long, and the ceiling was 15 m above the stage floor. The sound source was located 4.0 m from the front of the stage but was 0.5 m to one side of the centerline and 1.5 m above the stage floor. The front and rear walls were vertically bisected to obtain two faces, and each stretch wall along the side of the seating area was divided into four faces. The walls were kept vertical (i.e., tilting was not allowed) to examine only the plan of the hall in terms of maximizing $\overline{S_4}$, the preference scale values associated with the IACC, which may most effectively determine the shape of a room (Ando 2007 and 2014). Each wall was moved independently of the other walls. Forty-nine listening positions distributed throughout the seating area on a 2×4 m grid were selected. The moving range of each vertex was ± 2 m in the direction of the line normal to the surface. The coordinates of the two bottom vertices of each surface were encoded on the chromosomes for the GA. In this calculation, the most preferred listening level was set for a point on the hall's long axis (central line), 10 m from the source position. The result of optimizing the shape of the hall with $\overline{S_4}$ is shown in Fig. 13.2a. The rear wall of the stage and the rear wall of the audience area took on convex shapes.

For performers on stage, reflections in the median plane like above and in the rear are preferred (spatial aspect), and the delay time of reflection is greatly dependent on the effective duration of the running ACF, $(\tau_e)_{min}$ of the music source signals (temporal aspect) (Ando 1998).

A practical application of this design theory was realized on the Kirishima International Concert Hall (Miyama Conceru: Fig. 13.3, Photo 13.2, Ando 1998) and Tsuyama Music Cultural Hall (Belle Foeet Tsuyama: Photo 13.3, Fujii, Kato, Shimokura, Okamoto, Suzumura, and Ando 2004), which were characterized by a "leaf shape." In order to enhance individual satisfaction of each listener, a special facility in seat selection for testing each listener's own subjective preference was first introduced to Miyama Conceru (Ando and Noson 1997).

13.3 Design Study of an Opera House

The opera is an epitome of human life. When we see events occurring on stage that magically seem to be intertwined with our personal life, we feel as if we might be recipients of a special message forwarded to us through the drama, and through Nature as well. This message includes affections from the Sun, Moon, stars, waters, winds, clouds, trees, and flowers. These spontaneous interactions with Nature may act in favor of the flexibility of the opera. For example, occurring extraordinary

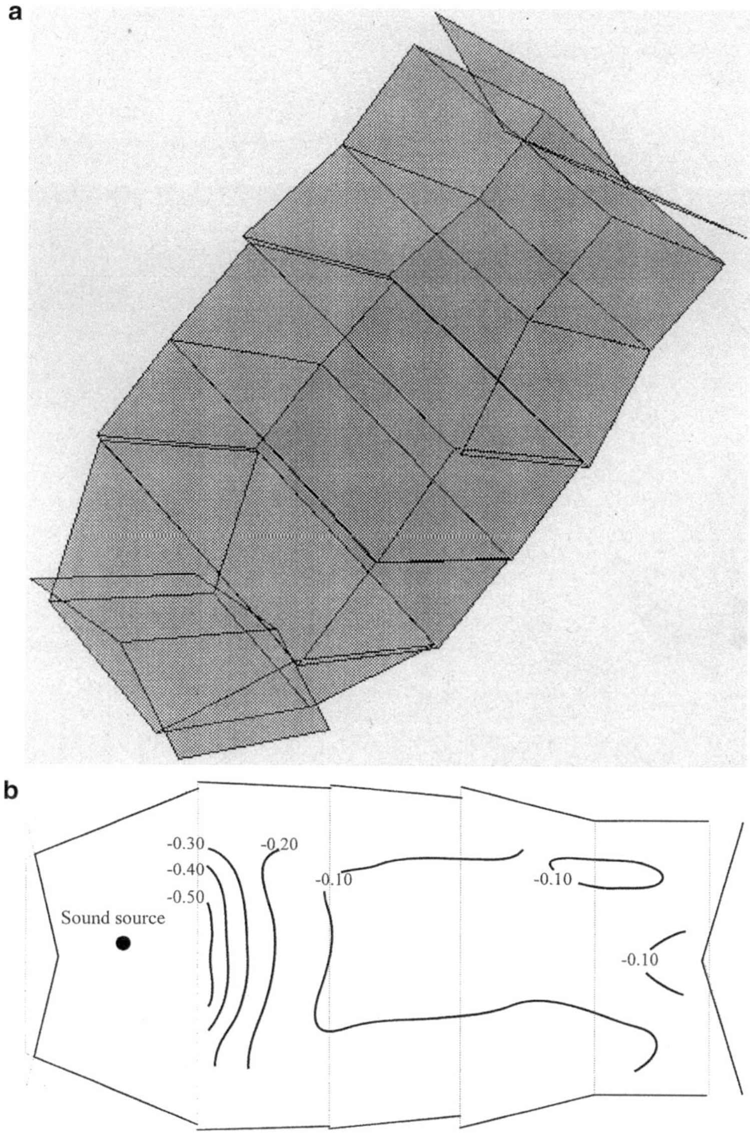

Fig. 13.2 (**a**) Resulting shape by the genetic algorithm (GA: Holland 1975) optimized with respect to the scale value of subjective preference due to the typical spatial factor, IACC. The rear wall of the stage and the rear wall of the audience area are convex shape. (**b**) Contour lines of equal scale value of subjective preference calculated for the geometry shown in (**a**)

weather changes may inspire individual creations that solve problems in art, science, or any other field of life. Since the time of the Big Bang, the environment, i.e., our Universe has undergone dramatic development with deep affections or a

b

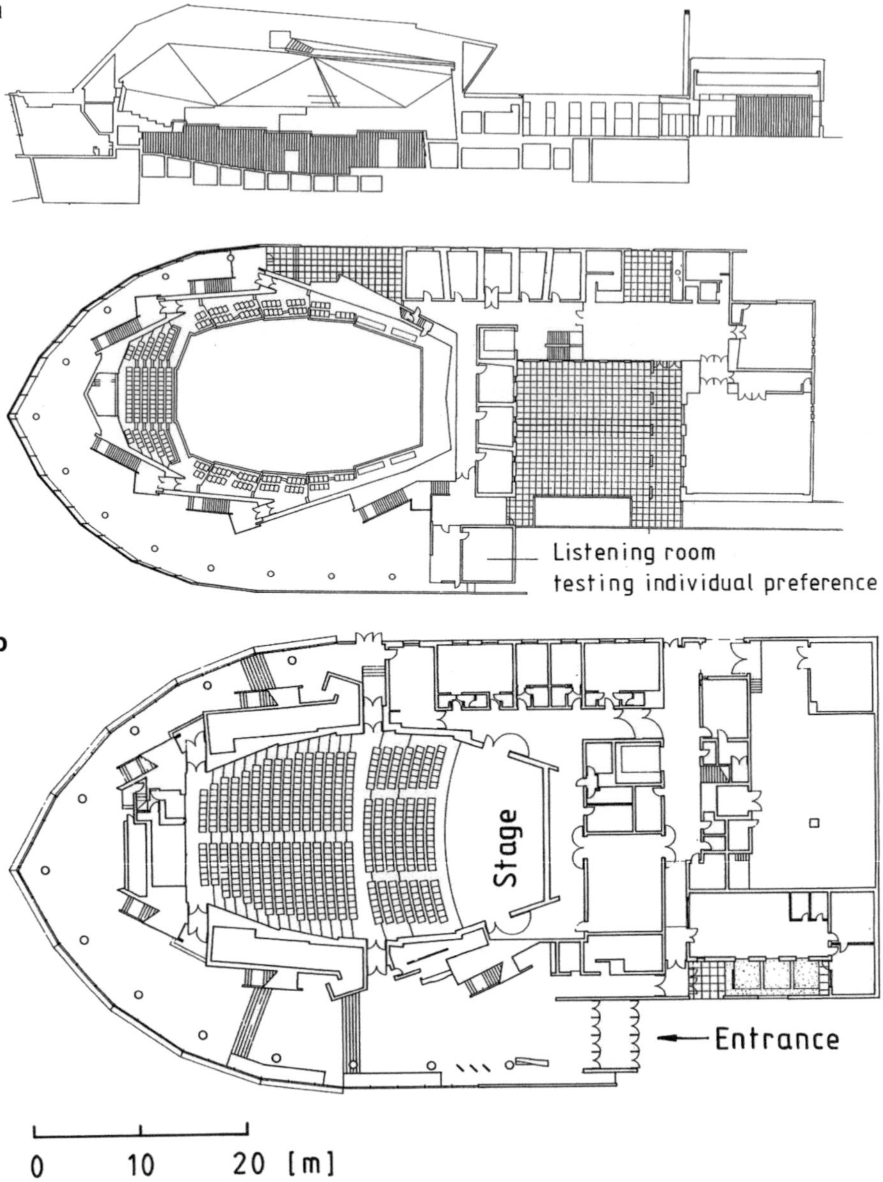

Fig. 13.3 Scheme of the Kirishima International Music Hall (Miyama Conceru) designed by Maki (1997). (**a**) Longitudinal section and plane of balcony level. (**b**) Plane of audience level

intentions from Nature that result as the phenomena that we experience at present stage. Thus, our life environment is truly surrounded by full affection from Nature, and we are receiving it always. Considering that our cyclic existence is part of the

Photo 13.2 Stage area of the Kirishima International Music Hall (Miyama Conceru)

longest period known, i.e., the cyclic universe (Appendix 3; Steinhardt 2009), *we may perceive that our environment is still performing many actions we are not yet capable of fully comprehending. It is not surprising why we may perceive some of them as magical. Stages of theaters blended with natural activities have been known as Japanese NOH theaters and in Javanese Gamelan theaters, as well as in ancient Greek and Roman theaters.*

In this section, a sketch of a design study for a four-seasonal opera house evoking individual creations is demonstrated. Through naturally participating in the opera, individuals might obtain inspiration, which may help them to discover their personalities and manifest creations accordingly.

13.3.1 Temporal Design

Firstly, at the design stage of an opera house, temporal design is carefully conducted (Ando 2009a, b, 2013; Ando and Cariani 2014). The discrete periods in Nature and the human body as shown in Fig. 8.1 are considered in the temporal design. The minimum period is in the order of 1 ms that corresponds to the neural firing rate, and the next distinct period is about 90 min, which relates to rapid eye

Photo 13.3 Tsuyoma Music Cultural Hall (Belle Foret Tsuyama) with 52 columns (each of the columns was 30 cm in diameter) located in front of the walls

movement (REM) associated with the brain rhythm of the waking state and the resting/sleeping state. This is also an indication of why we need and should have a short rest between programs of opera and drama. The period of 1 day is the most fundamental in life.

Possibilities of design applications are literally endless, so let us start with an early morning opera beginning before sunrise, continue with the usual opera beginning in the evening at 7:00 pm, and finish the day with a sunset opera. If different programs of opera and drama performance are planned for every weekend, then no additional costs for public notifications are required because of the periodicity of the performance schedule. The opera may inspire other possible programs such as the "moonlight opera," and the "opera of the four seasons" which could be planned for the full moon each month and each year, respectively. The concept of a "New Year opera" at sunrise would tie human activity with cosmic events.

Although somewhat common, it is worth noting that human life is usually considered a combination of the first stage of the body and the second stage of the mind. However, for a more joyful and peaceful life, it is necessary to introduce the third stage of personality-oriented creations based on the individual "seed" (Fig. P.4). This is what the audience as well as the performers may really enjoy.

Allowing more varieties of human life to be blended with opera performances and drama means creating the temporal and the spatial environment for it that would stimulate the left and right human cerebral hemispheres, respectively. These three stages of human life, the temporal and spatial design may all be considered in actual life as well as in opera performance. The temporal design of opera houses and human life in general is important for further possible creations of opera, drama, and other productions, so that a more fruitful cultural life may result.

13.3.2 Crystal Opera House

According to the plot of a given opera or drama being performed, the ceiling together with the upper roof as well as the rear walls behind the audience floor could be opened, whereas the side walls are in fact strengthened double-glazed windows. For example, audiences may look at shooting and twinkling stars, imagining a story of the Universe taking place at the same time. In certain situations, a breeze of fresh air from the outside relative to an entirely closed enclosure may make us feel like we are receiving deep affection from Nature. Another merit of this kind of opera house is that it allows people walking outside to witness the performances taking place inside and make them want to join the audience in the house (Ando 2015).

Due to the temporal design and the preference theory of sound fields, a possible new type of crystal opera house may be proposed as shown in the sketches of Fig. 13.4a, b. In order to stimulate creation in such an opera house, and to attain an idea of the full variety of possible performances, three different stages, i.e., inner stage, upper stage, and outer stage of the opera house together with Nature may be fully utilized.

The rear wall of the stage is a large glass looking over the outer stage and farther scenery, overlooking the sea as indicated in the sketches. Similar large glasses are utilized for side walls as well, blending performances with natural activities. A pit elevator is useful for adjusting the level of the floor for orchestra performance and to enlarge the inner stage.

Such an opera house could encourage further possible creations of drama and opera.

13.4 Open Space for Performances

A typical open space is a courtyard with a superior sound field as shown in Fig. 13.5a that was first noticed by Bekesy (1934, 1967). Figure 13.5b indicates the most recommended position in this courtyard which is marked by $[S]_p$ (Ando

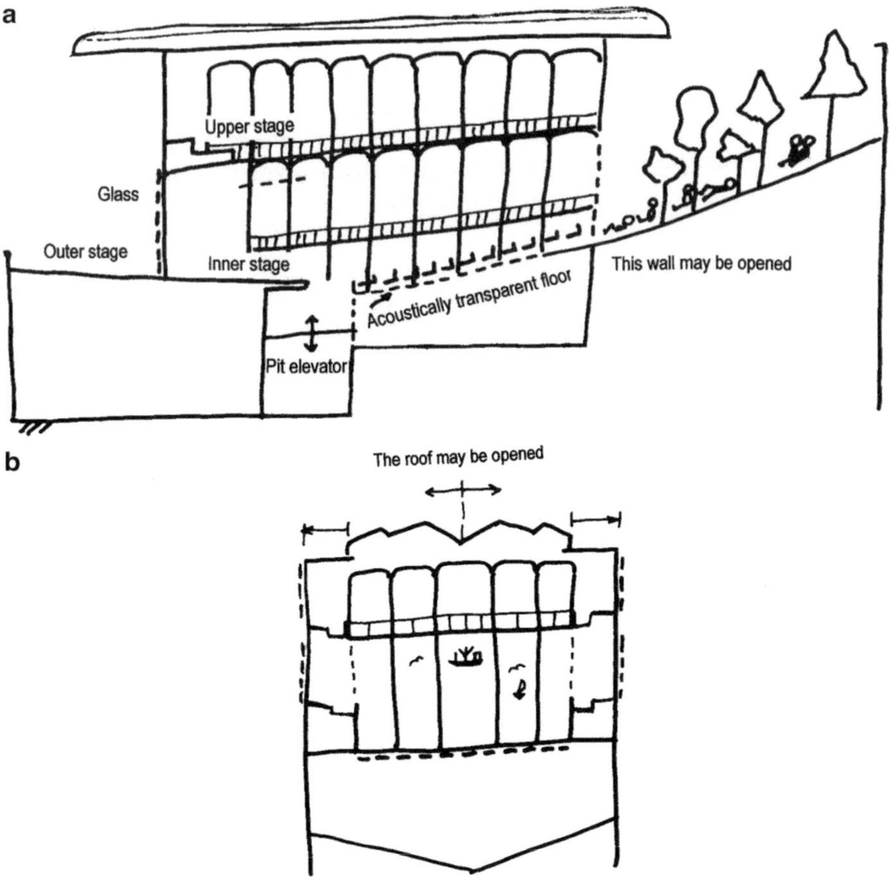

Fig. 13.4 (**a**), (**b**) An example of cross-sectional sketches for the opera house proposed hopefully to stimulate individual creations. Walls are glass walls that enable better blending with Nature. For example, a possible plot for an opera could involve utilizing the three different stages for finding the protagonist's mission in life, particularly his or her individual potentiality for creations. The story could also be about finding preferred environments for the third stage of human life to approach peace, avoid bullying or discrimination, which are all things that epitomize humanity and set humanity apart from animal life, which only involves the first and second stages of life. Acoustically transparent floors help to avoid large dips in the low-frequency range between 100 and 500 Hz due to interference between the direct sound and the reflection from the floor (Takatsu et al. 2000)

1998). These contour lines of equal scale value of subjective preference were calculated at the source position for music (Motif B). When music or drama is performed, people may enjoy it from the windows of their houses as well as by sitting in the courtyard.

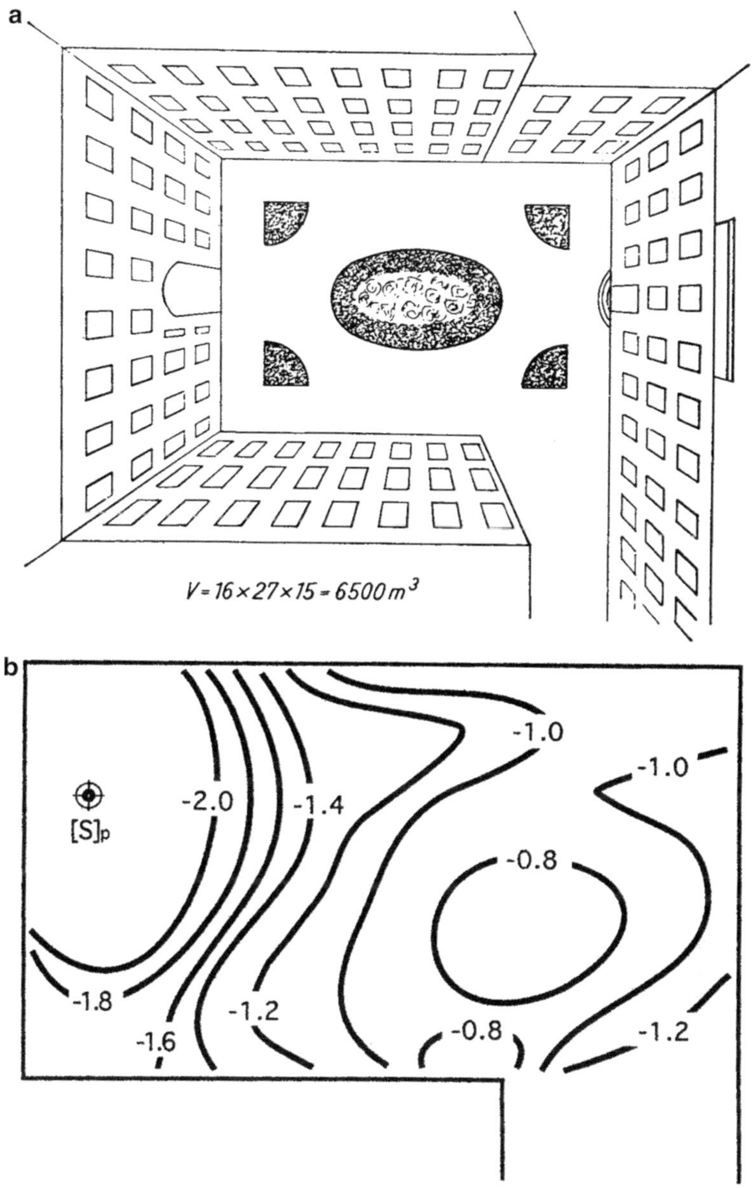

Fig. 13.5 (**a**) A courtyard with a superior sound field (Bekesy 1934, 1967). (**b**) When the performing position is located at the most effective source location, *contour lines* of equal scale values of subjective preference for listeners are calculated by Eq. (5.2) (Ando 1998)

This kind of courtyard is also enjoyable as a play area for children who like receiving reflections of their own voices from the surfaces. The parents may also keep an eye on their children through the windows.

References

Ando Y (1983) Calculation of subjective preference at each seat in a concert hall. J Acoust Soc Am 74:873–887

Ando Y (1985) Concert hall acoustics. Springer, Heidelberg

Ando Y (1998) Architectural acoustics, blending sound sources, sound fields, and listeners. AIP Press/Springer, New York

Ando Y (2004) On the temporal design of environments. J Tempor Des Architect Environ 4:2–14. http://www.jtdweb.org/journal/

Ando Y (2007 and 2014) Concert hall acoustics based on subjective preference theory. In: Rossing TD (ed) Springer handbook of acoustics. Springer, New York

Ando Y (2009a) Theory of temporal and spatial environmental design. In: McGraw-Hill yearbook of science & technology 2009. McGraw-Hill, New York, pp 384–389

Ando Y (2009b) Auditory and visual sensations. Springer, New York

Ando Y (2013) Environmental design for the third stage of human life (persistence of individual creations). J Tempor Des Architect Environ 12:1–12. http://www.jtdweb.org/

Ando Y (2015) Opera house acoustics based on subjective preference theory. Springer, Tokyo

Ando Y, Cariani P (2014) Neurally based acoustics and visual design. In: Kiang N, Sessler GM (eds) Acoustis, information, and communication. Springer, Cham

Ando Y, Noson D (eds) (1997) Music and concert hall acoustics, conference proceedings of MCHA 1995. Academic, London

Bekesy G (1934) Ueber die Hoersamkeit von Konzert und Rundfunksaelen. Eleckrische Nachrichten Technik 11:369–375

Bekesy G (1967) Sensory inhibition. Princeton University Press, Princeton

Holland JH (1975) Adaptation in natural and artificial systems. The University of Michigan Press, Ann Arbor

Maki F (1997) Sound and figure: concert hall design. In: Ando N (ed) Music and concert hall acoustics, Conference proceedings from MCHA 1995

Sperry RW (1974) Lateral specialization in the surgically separated hemispheres. In: Schmitt FO, Worden FC (eds) The neurosciences: third study program. MIT Press, Cambridge. Chapter 1

Steinhardt PJ (2009) Cyclic universe theory. McGraw-Hill Yearbook of Science & Technology, New York, pp 69–71

Takatsu A, Hase S, Sakai H, Sato S, Ando Y (2000) Acoustical design and measurement of a circular hall, improving both spatial and temporal factors at each seat. J Sound Vib 232:263–273

Chapter 14
Theory of Environmental Design and Escape from Vandalism

14.1 Theory of Environmental Design Incorporating Temporal and Spatial Values

It is assumed that Eq. (5.2) for the sound field may be applied to all physical environments, such that

$$\begin{aligned} S = & [g_{\mathrm{L}}(x) + g_{\mathrm{R}}(x)]_{\text{Sound field}} \\ & + [g_L(x) + g_R(x)]_{\text{Visual field}} \\ & + [g_L(x) + g_R(x)]_{\text{Thermal field}} \end{aligned} \tag{14.1}$$

where $g_{\mathrm{L}}(x)$ and $g_{\mathrm{R}}(x)$ are the scale values of the temporal and spatial factors in the left and right hemisphere processing, respectively. In particular, this is true at least in the neighborhood of the optimal conditions of each physical factor, avoiding extreme physical environments such as listening to music in temperatures below 0 °C. It is an expressed hope that scientific investigations for the thermal field as well as the vibration field will be conducted with respect to the temporal and spatial factors.

In architectural planning and design, the temporal factor associated with the left hemisphere has been widely ignored although it is a dimension most inherent in our own makeup, in both physical body as well as our consciousness. As shown in Fig. 8.1, the crucial factor in the temporal dimension of the environment is the periodic cycle. For example, the shortest period (about 0.5–5 s) is related to brain waves, which are associated with temporal percepts of sound, visual and thermal fields, such as music, the twinkling stars, and temperature change. The rapid eye movement (REM) of about 70–150 min, related to the basic rest-activity cycle, is associated with, for example, one session of a concert, lecture, and work. The circadian rhythm deeply connected to sunlight and the Earth's rotation period is associated with daily human activity. The week created by social custom for work and leisure is associated with, for example, the planning period of concert and

Y. Ando, *Brain-Grounded Theory of Temporal and Spatial Design*,
DOI 10.1007/978-4-431-55891-0_14

drama or social activity without additional public notification costs. The next Eigen period is concerned with the movement of the Moon. The revolution of the Earth around the Sun is associated with the color change of leaves and annual festivals. The black spots on the Sun that appear about every 11 years may influence environmental conditions on Earth. The alternation of generations every 30 years and the natural human life span of roughly 90–100 years should be considered in the planning of houses in accordance with the individual purpose or mission of life. There are other longer cosmic periods in addition to the ones mentioned herein.

The present theory suggests that these discrete periods be explicitly recognized during the design process for any human activity. The passage of time in the designed environment should be as consciously considered as the three-dimensional organization of space itself. There are specific dimensions of space such as a room, house, building, region, urban area, and so on. It is highly recommended that all matrix elements representing the discrete periods and space ranks shown in Figs. 9.1 through 9.3 and 9.5 be considered in the design process.

14.2 Indication of Quick Escape from Dangerous Occurrences

Seeing the global climate change, the temporal and spatial design of environments should be carefully considered. In this section, effective indications for quick and safe escape from dangerous occurrences are discussed.

It is known that the right hemisphere is specialized in developing quick spatial percepts such as pattern recognition, face identification, and percepts of textures as discussed in Chaps. 2 and 3. Such spatial information is processed in the right cerebral hemisphere, which is directly connected to human action or operation without time lag, enabling escape from dangerous occurrences.

On the other hand, the left hemisphere is specialized in temporal percepts or sequential processing of information such as speech recognition. That is to say, understanding continuous speech takes a relatively long time, and thus indication of speech is not effective for quick action to escape from dangers.

Any information containing nonverbal indications is displayed to the right hemisphere, so that we can immediately escape from fire, for example. One effective indication to an escape exit is an arrow chain with a red on-and-off light (luminescence) in any building or vehicle. The color red is the most effective for use in smoke, due to the fact that it has the longest visible wavelength.

A standing signboard shown in Photo 14.1 saying "There is a difference in the level of way, please watch your step" was of no use, because such information is processed in the left hemisphere, and people do not care for words in such situations. Consequently, many concert goers tripped and fell on the ground and some even got injuries instead of enjoying the passage of time.

Photo 14.1 There is an indication of notifying difference in the level of way. "Please watch your step." But it was of no use, because people do not care for words in such situations. Steps should be in different colors for display to the *right* hemisphere

Another example with a more tragic end occurred when a student taking military flying lessons took off an airport on a plane together with the instructor's plane. At an altitude of about 10 km, the jet engines of the student's plane suddenly stopped running. The instructor said to the student that they have plenty of time to restart the engines and asked the student to find a red knob under the pilot seat and pull it. But the student answered that there was no knob under the seat in his machine. Clearly, in a dark environment, it would be hard to find one. The instructor repeated his advice over and over again. Unfortunately, the student's plane crashed and he did not survive. Had the knob been programmed to automatically self-indicate with on-and-off light, it would have been easily found without any conversation.

Another tragic example is from a store where a bargain sale was going on. A loudspeaker system suddenly announced that the lower floor had caught fire and all shoppers should exit the building immediately. However, nobody escaped from the burning building, because speech is processed in the left hemisphere, and the verbal information was not applicable to the hazardous situation. Unfortunately, many people died in the fire.

14.3 Escape from Effects of Aircraft Noise on Developing Children

Integrated effects of noise on the development of children are mentioned to show why it is recommended to escape noisy living areas.

Environmental noise is a representative measure of quality of environment in urban life. Our brain vividly and instantaneously responds to changes in the environment within a period of less than about 1.0 s. The auditory-temporal window (2 T) is such a short period (Ando 1998). In other words, temporal sensations that change very slowly over a longer period (more than several minutes) are in fact very weak in effect on our brain. However, we should be aware of the "long-term integrated effects" that environmental noise has on unborn babies and children who are exposed to noise over the course of more than 1 year.

Since 1968, investigations have been conducted around the Osaka International Airport in Japan. Results clearly indicate:

1. Effects on the body: Integrated effects of aircraft noise in a living area (Fig. 14.1) have been described in terms of human placental lactogen (HPL) as shown in Figs. 14.2 and 14.3 (Ando and Hattori 1977b). Moreover, the development of unborn babies as demonstrated by the number of low-birth-weight babies in relation to the number of jet planes is shown in Figs. 14.4 and 14.5. (Ando 1988). After birth, an effect on the development of height has been discussed (Schell and Ando 1991).

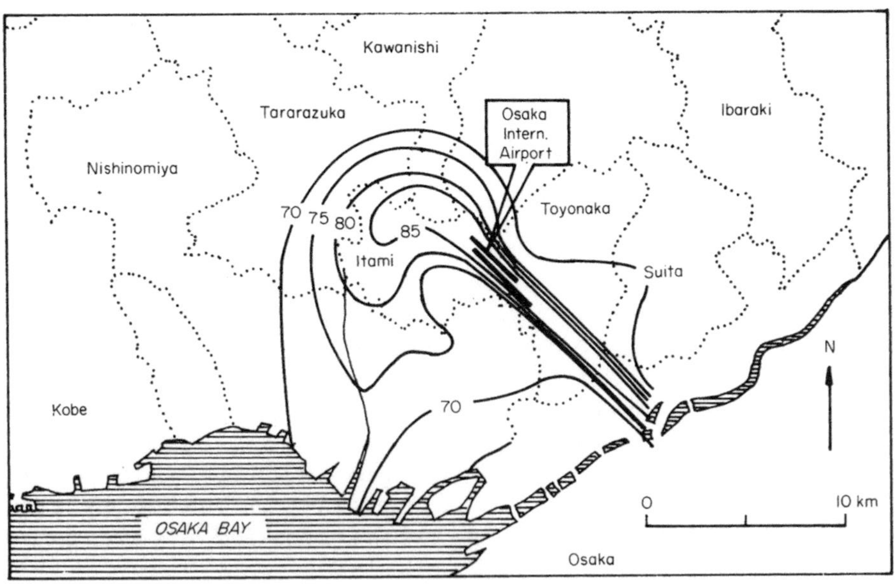

Fig. 14.1 Contour lines of equal WECPNL around the Osaka International Airport. WECPNL in most of the area of Itami City was in the range of 85 < WCPNL < 70

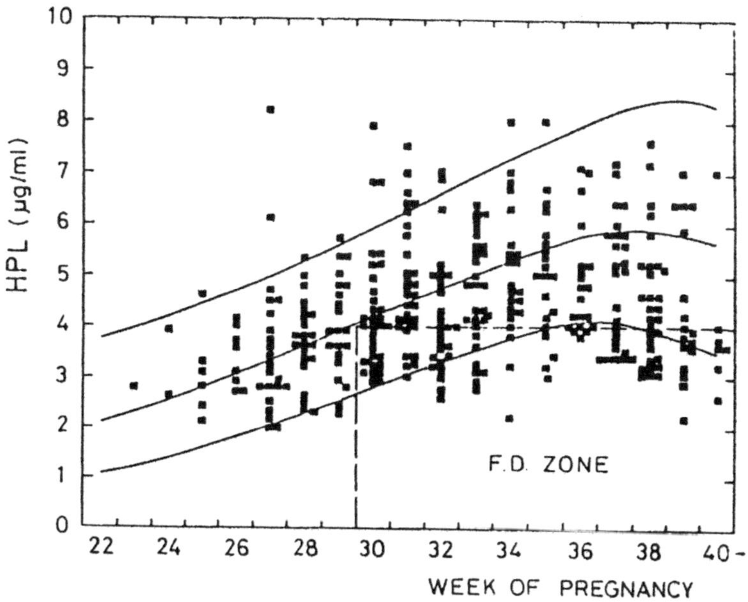

Fig. 14.2 HPL levels of 343 mothers living in Itami City (Ando and Hattori 1977a, b). The *three lines* show the mean and ±2SD values for normal HPL levels from Lindberg and Nilsson (1973)

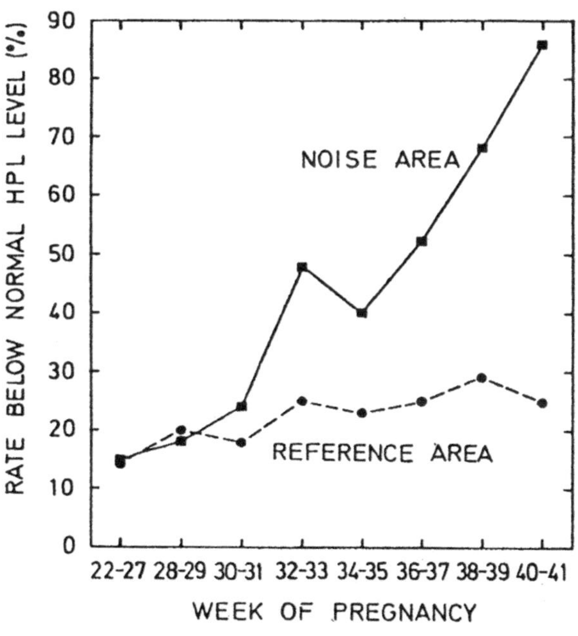

Fig. 14.3 Percentage of subjects with HPL levels more than 1SD below the mean by stage of pregnancy (Ando and Hattori 1977b)

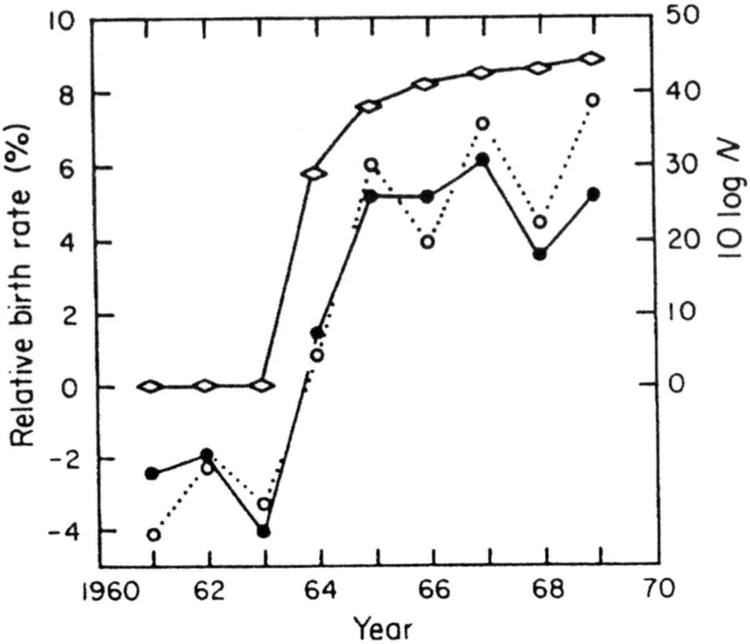

Fig. 14.4 Birth rates below 3000 g as a function of year of living in Itami City, in reference to neighboring cities without aircraft noise shown in Fig. 14.1. ●: male. ○: female. Rhomb: number N of aircrafts flew over Itami City plotted in a scale of 10logN

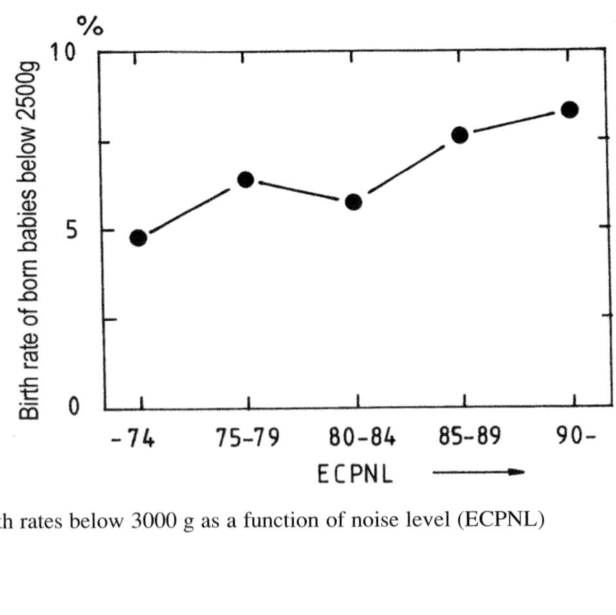

Fig. 14.5 Birth rates below 3000 g as a function of noise level (ECPNL)

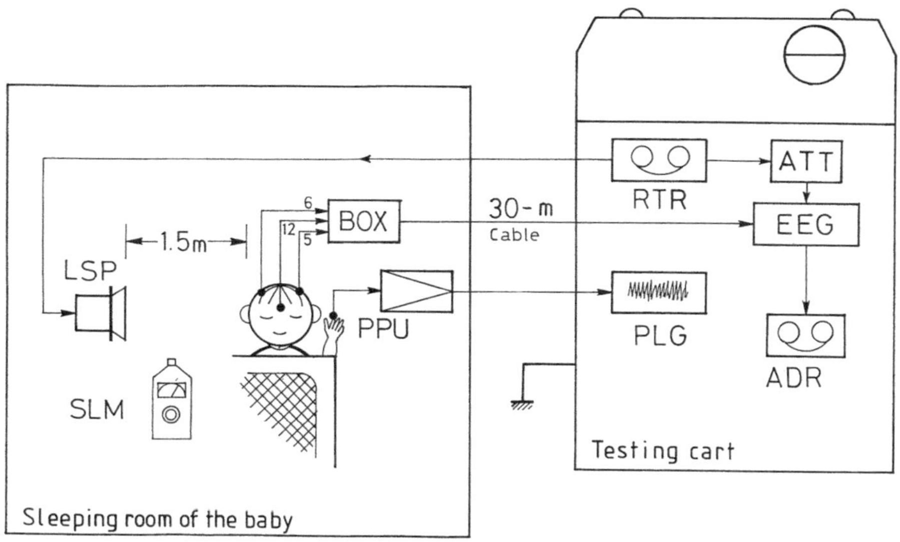

Fig. 14.6 Recording system of plethysmogram (PLG) and EEG of sleeping babies in their actual rooms

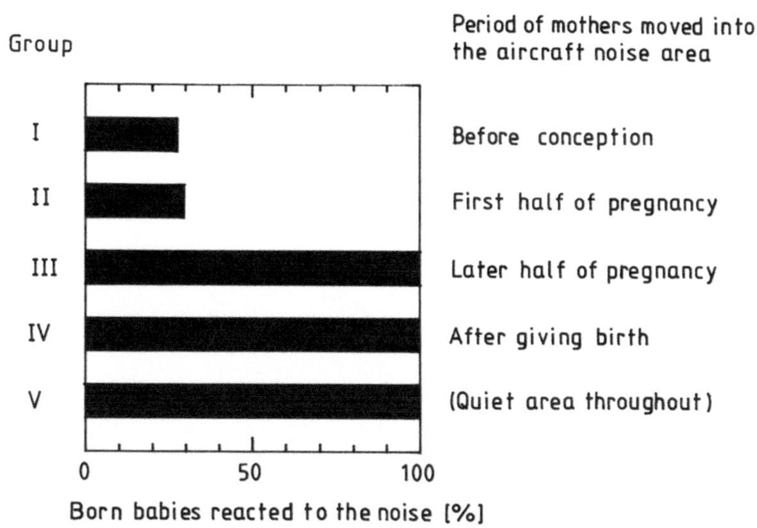

Fig. 14.7 Reactions to the noise below 80 dBA recorded by PLG for each group. When mothers had lived in the noise area from the first half of pregnancy, about 70 % of babies born of mothers did not react during sleep even when the jet plane noise was 80 dBA

2. Effects on the mind: Postnatal effects of aircraft noise on the sleep of newborns are dependent on the period during which their mothers entered the noise area in reference to the period of pregnancy as shown in Figs. 14.6 and 14.7 (Ando and Hattori 1977a). Figure 14.8a, b display how the development of hemispheric

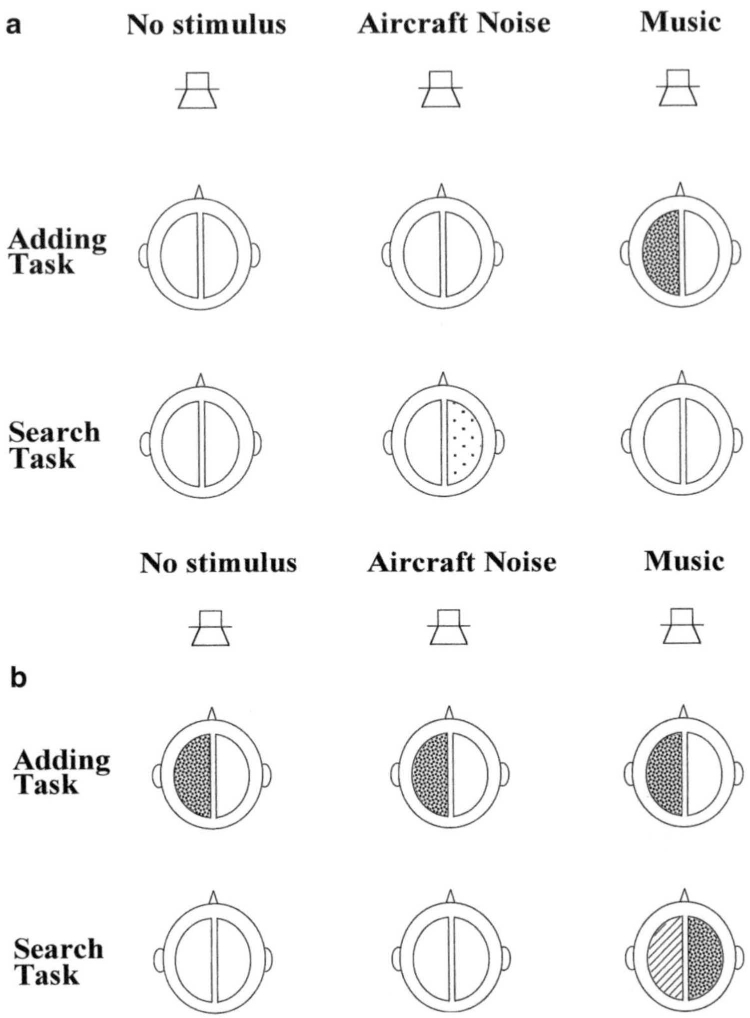

Fig. 14.8 (**a**) Explanations of interference between mental tasks and sound stimuli for children from *a quiet living area*. Interference effects shown by *shaded areas* differ remarkably from noisy and quiet living areas. The hemispheres marked were disordered due to a sudden drop of the mental tasks. According to the auditory-brain system model, noise and music are, respectively, associated with the *right* hemisphere and the *left* hemisphere. And the adding task and search task, respectively, are associated with the *left* hemisphere and the *right* hemisphere. (**b**) Explanations of interference between mental tasks and sound stimuli for children from *a noisy living area*. The hemispheres marked were disordered due to a sudden drop of mental tasks. According to the auditory-brain system model, for quiet areas without aircraft noise, noise and music are, respectively, associated with the *right* hemisphere and the *left* hemisphere. And the adding task and search task are associated with the *left* hemisphere and the *right* hemisphere, respectively

specialization in children over long-period noise was clearly described by results from testing two different types of mental work associated with the left and the right hemispheres (Ando et al. 1975; Ando 1988).

Effects on creation: It is not clear yet how environmental noise affects creativity; however, the integrated effects that environmental noise has on the specialization of cerebral hemispheres as listed above, in turn, may influence the development of the personality as the source of creation. May a poem read by Otomo no Yakamochi on 3 April, 653, collected in the volume "Manyo-Shu" (famous eighth century poetry) from the oldest Japanese anthology of poems inspire the reader: "From a patch of bamboo trees in the garden of my house, a wind in this evening might produce a faintly sound." This demonstrates that a quiet environment where leaves of bamboo trees rustle in the wind can concentrate into human creations.

14.4 Sleep Disturbance by Upstairs-Toilet Noise with Changing Spatial Factors

The purpose of this study was to identify factors of the flushing noise of an upstairs toilet, which despite having a very low SPL (<36 dBA at the peak) caused annoyance and sleep disturbance for an apartment resident. The temporal and spatial factors of the flushing noise from an upstairs toilet to the head position on a bed were analyzed because the resident suffered from sleep disturbance. Results of IACC and τ_{IACC} *as a function of time indicated that the flushing noise was rotating around the head dynamically in addition to the large change in temporal factors including the SPL. These facts indicate that the left and right hemispheres were simultaneously stimulated, so that even though the sound level was very low, the resident was annoyed and suffered from sleep disturbance.*

The noise signals were picked up by 1/2-in. condenser microphones placed at two ear entrances on a spherical dummy head placed on the bed. The measurement time for the upstairs flushing toilet noise was 5 s (Fig. 14.9). The values of all temporal ACF factors were obtained with the integration interval of 0.5 s and the running interval of 100 ms. As shown in Fig. 14.10a–d, the SPL measured by $\Phi_{ll}(0)$ for the flushing toilet noise extracted from the running ACF was 29–35 dBA, and the background noise was less than about 23 dBA (Kitamura et al. 2002). Thus, the maximum signal-to-noise ratio was 12 dBA. If the difference between the background noise level and the noise signal level is greater than 10 dB, the background noise does not significantly affect the noise signal measurement (Beranek 1971). The τ_e value for the flushing toilet noise exceeded 100 ms with $\phi_1 > 0.5$ while $\tau_e < 0.1$ ms and $\phi_1 < 0.01$ of the background noise throughout the measurement time. Thus, the flushing toilet noise had much more repetitive features than the background noise. It has been reported that loudness increases in proportion to the value τ_e (Merthayasa and Ando 1996). The value of τ_e for the flushing toilet noise was the largest near the peak of $\Phi_{ll}(0)$. The value of τ_1 for the flushing

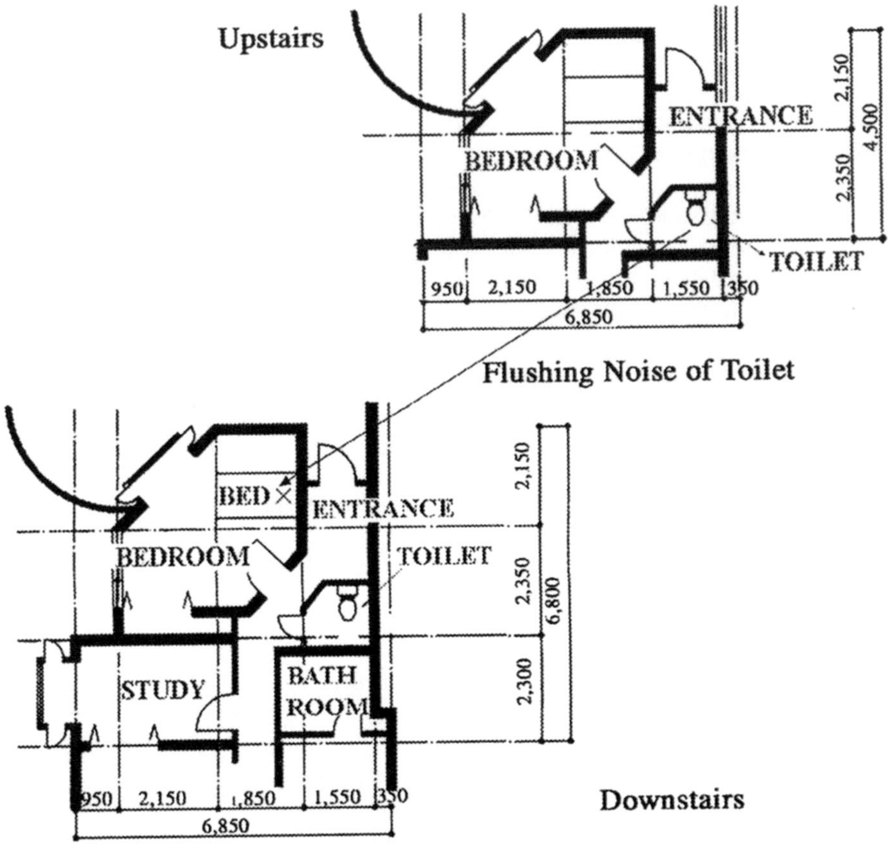

Fig. 14.9 Plans of upstairs and downstairs in an apartment. The flushing noise of an upstairs toilet was recorded on a bed in a hexagonal room downstairs

toilet noise had a discrete value at 3.6 ms, which means that the perceived pitch was 275 Hz. The background noise did not have any clear pitch and tonal components, similar to white noise.

In addition, the measured spatial factors extracted from the running IACF are shown in Fig. 14.11a–c. The IACC value for the flushing toilet noise was much higher than for the background noise (below 0.05) in the measurement time. Thus, no specific directions could be perceived for the background noise. On the contrary, all spatial factors, the IACC, τ_{IACC}, and W_{IACC} of the flushing noise, changed dramatically as a function of time. These results signify that subjective diffuseness, localization of the sound source, and the ASW of the flushing toilet noise changed greatly as a function of time. Particularly, values of τ_{IACC} varied from −0.6 to +0.6 signifying that the noise was perceived as rotating around the head of the sleeper. This phenomenon could be caused by the hexagonal shape of the bedroom.

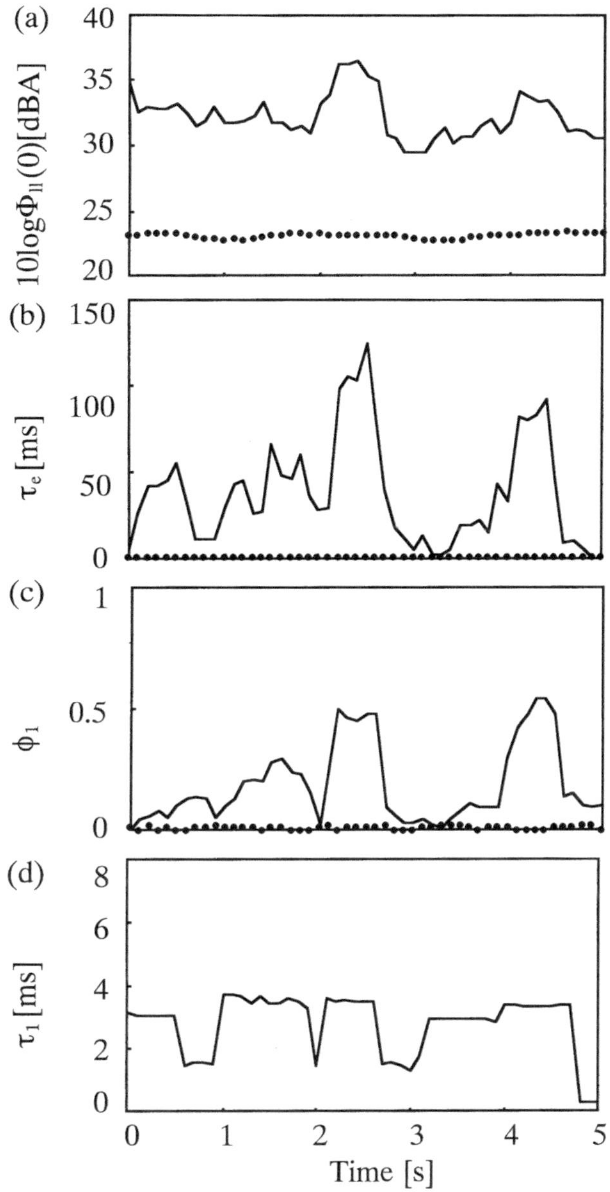

Fig. 14.10 Measured factors extracted from the running ACF. *Solid line* indicates values for the flushing toilet noise and *dotted line* indicates the background noise. (**a**) $\Phi_{ll}(0)$. (**b**) τ_e. (**c**) ϕ_1. (**d**) τ_1. The values of all factors were obtained every 100 ms with an integration interval of 0.5 s

Thus far, it has been shown that the temporal and spatial factors extracted from the ACF and IACF of the flushing toilet noise had specific characteristics. These facts imply that both temporal percepts and spatial percepts of the flushing toilet noise changed dramatically. According to the auditory signal processing model (Fig. 2.11: Ando 1985, 2009), temporal information is mainly processed in the left hemisphere and spatial information is mainly processed in the right hemisphere

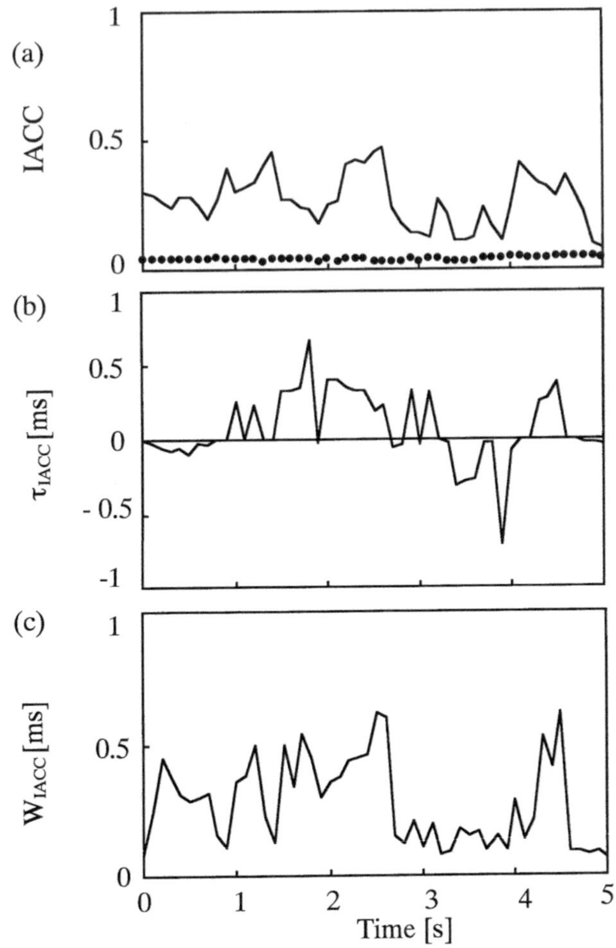

Fig. 14.11 Measured factors extracted from the running IACF. *Solid line* indicates values for the flushing toilet noise and *dotted line* indicates the background noise. (**a**) IACC. (**b**) τ_{IACC}. (**c**) W_{IACC}. The values of all factors were obtained every 100 ms with an integration interval of 0.5 s

(Ando 1998). Thus, the flushing noise from an upstairs toilet may simultaneously stimulate both the left and right hemispheres of this resident. This may explain why the resident felt that the flushing noise of the upstairs toilet was very annoying and disturbed his sleep despite its very low SPL.

14.5 Escaping Ill Social Groups and Preserving the Third Stage of Life

As mentioned earlier, daily environmental noise affects the specialization of cerebral hemispheres in children. Loud majority opinions, in turn, usually result in an individual losing their sense of preference which is key to maintaining their own

life. Consequently, individuals who, before finding their unique personalities, are subject to blind following or belong to special groups with mass psychology or "magic", are at great risk of losing their mental and physical health, not to mention the third stage of their life. There are, for example, newly-risen religions without any affection toward living creatures, companies targeted at only making money without any culture, violent military countries, and cults that engage in (sexual) unions under the auspices of "the divine". For the purpose of preserving the third stage of life, the most unique expression of creativity of any human, individuals should safeguard themselves from such ill social groups.

Every moment of time is precious, and no temporal unit should be slighted in the search for individual personality that would integrate possible creations, which may contribute to social culture and the preservation of life. From unique DNA code to unique environments, a rare melding of personal preferences forms a personality. Starting from the very beginning of life, parents could promote the development of their child's rare talent by providing a temporally and spatially preferred environment including facilities like the CWS (Chap. 11).

On an evening in March in 1952 after the death of his father, the author entertained some apprehensions about the future. After a while, it started to get dark and he returned home but nobody was in the house. He happened to find the Moon throwing light on the tatami floor through a simple window as shown in Drawing 14.1. In this moment, he perceived deep affection streaming from the universe and woke up to a realization of the fact that the universe is always with him.

Another memorable occurrence took place when a forest on a hill next to a living area had caught fire. Looking at the fire, his mother said to him that if this kind of fire continued to other regions on Earth, then the environment would not be good enough to sustain life. This was the first motivation to start investigating preferred environments. She continued her hypothesis by saying that this world had died and begun again repeatedly. This hypothesis is still not solved and remains in the author's mind. These defining moments were the beginning for finding the third stage of life.

Set A shown in Fig. P.4 is limited, and set A^C is reconfirmed as infinity, thus waiting to be solved by individuals for the sake of themselves and others.The third stage of life is an escape from unhealthy (social) conditions to a life unique driven by a mission.

In order to assist developing personality as shown in Fig. P.4 in Preface, Table 14.1 lists typical examples realizing temporal design based on preference theory for all of the three stages of human life aviding serious time-integrated effets of noise on human developments.

Concluding remarks of this volume are:

1. The ill concept of "time is money" leads to economical competition and thus ill treatment of others at the individual level.
2. Consequently, wars and environmental disruption driven by economical reasons occur.

Drawing 14.1 Moonlight on a tatami floor, a typical sign of affection from the universe

3. However, if individuals know the special talent offered to them by their personality that stems from their unique DNA and a preferred environment is provided so that their personality can develop, then creations that will integrate and become culture will live for a long time after the end of the first (body) and second (mind) lives, constituting the third stage of life.
4. The new concept of the third stage of life is fundamental to maintaining the environment and establishing lasting peace on Earth.

Table 14.1 Typical examples of temporal design for the three stages of human life

Three stages of human life		Environment to be avoided	Environments to be designed
1	**Body**	(1) Time-integrated effects of noise on unborn babies	
2	**Mind**	(2) Time-integrated effects of noise on brain development	(3) Concert hall acoustics design based on auditory-brain model
			(4) Seat selection enhancing individual satisfaction, maximizing individual preference
3	**Creation based on personality**		(5) Creative work space (CWS) for the left and right hemispheric tasks

References

Ando Y (1985) Concert hall acoustics. Springer, Heidelberg

Ando Y (1988) Effects of daily noise on fetuses and cerebral hemisphere specialization in children. J Sound Vib 127:411–417

Ando Y (1998) Architectural acoustics, blending sound sources, sound fields. Springer, New York

Ando Y (2009) Auditory and visual sensations. Springer, New York

Ando Y, Hattori H (1977a) Effects of noise on sleep of babies. J Acoust Soc Am 62:199–204

Ando Y, Hattori H (1977b) Effects of noise on human placental lactogen (HPL) levels in maternal plasma. Br J Obstet Gynaecol 84:115–118

Ando Y, Nakane Y, Egawa J (1975) Effects of aircraft noise on the mental work of pupils. J Sound Vib 43:683–691

Beranek LL (1971) Noise and vibration control. McGraw-Hill, Inc, New York

Kitamura T, Shimokura R, Sato S, Ando Y (2002) Measurement of temporal and spatial factors of a flushing toilet noise in a downstairs bedroom. J Tempor Des Architect Environ 2:13–19. http://www.jtdweb.org/

Lindberg BS, Nilsson BA (1973) Variations in maternal plasma level of human lactogen (HPL) in normal. J Obstet Gynaecol Br Commonw 80(7):619–626

Merthayasa I Gde N, Ando Y (1996) Variation in the autocorrelation function of narrow band noises: their effect on loudness judgment. In: Japan and Sweden symposium on medical effects of noise

Schell LM, Ando Y (1991) Postnatal growth of children in relation to noise from Osaka International Airport. J Sound Vib 151:371–382

Chapter 15
Remarks

15.1 Remarks on Personality-Oriented Preference and Creations

Preference is a joint response of the left and right cerebral hemispheres associated with time and space, respectively, that are stimulated by different signals during the three stages of life.

Animals have a body and a mind, but not the third stage of life, which is distinctive of humans and made up of creations and integrations. Manifest creations are based on an individual's unique DNA "seed". An individual's purpose of life or mission is given by Nature and it is to contribute to maintaining mankind together with other living creatures, all the while minimizing personal stress for the benefit and emergence of healthy creations. A lasting peace may be achieved by release of each personality given by Nature. Note that personal preference is quite different from egoism or greed.

One should not kill the human body and/or mind, nor should one plagiarize creations or ideas produced by other individuals. These are all lives, i.e., first, second, and third lives, respectively. In particular, the potential for the third stage of life exists for every individual even up to the last moment of the first and second lives, waiting to be unfolded to contribute to sustaining mankind and establishing lasting peace.

15.1.1 Remarks on Subjective Preference

Subjective preferences, being the evaluative judgments that actually steer an organism's behavior so that it may survive and propagate, are therefore the most primitive responses in organisms. Preferences guide the organism in the direction of maintaining life, thus we may observe brain activities related to preferences, which

Y. Ando, *Brain-Grounded Theory of Temporal and Spatial Design*,
DOI 10.1007/978-4-431-55891-0_15

are simultaneously deeply related to the aesthetic sense on an individual and on a global scale. The paired-comparison method has been used to obtain scale values of preference to be linked with corresponding brain activities. Acoustic design of a concert hall where preference scale values are maximized is a typical non-visible environment that forms the majority of a physical environment. It is, therefore, a good example for demonstrating the design process. In addition, an opera house is a kind of representation, a reduced environment of our real lives, as "All the world's a stage," and we are all on it. The theory of subjective preference for the sound field is described by the use of the temporal and spatial factors extracted from the respective correlation function mechanisms existing in the neural system (Ando 1985, 1998, 2007, 2009a). The specialization of the cerebral hemispheres may be significant in the formation of the independent effects that temporal and spatial factors evoke in a subject. Finally, a human-centered theory for planning the environment, so that it incorporates temporal and spatial factors for the left and right cerebral hemispheres, respectively, is proposed (Ando 2009b).

15.1.2 Remarks on the Theory of Subjective Preference (Chap. 5) for Designing the Environment

1. Subjective preferences are the most primitive responses.
2. Preferences are evaluative judgments.
3. Preferences guide the organism in the direction of maintaining life; thus, we may observe brain activities related to preference (Chaps. 2 and 3).
4. Subjective preference is deeply related to an individual and global aesthetic sense.
5. Acoustic design of a concert hall where preference scale values are maximized at each seating position is a typical non-visible environment that forms the majority of a physical environment.
6. The theory of subjective preference for the sound field is described by use of the temporal and spatial factors extracted from the respective correlation function mechanisms existing in the neural system.
7. The specialization of the cerebral hemispheres may be significant in the formation of the independent effects that temporal and spatial factors evoke in a subject.
8. In this volume, a human-centered theory for planning the environment so that it incorporates temporal and spatial factors for the left and right cerebral hemispheres, respectively, is proposed.

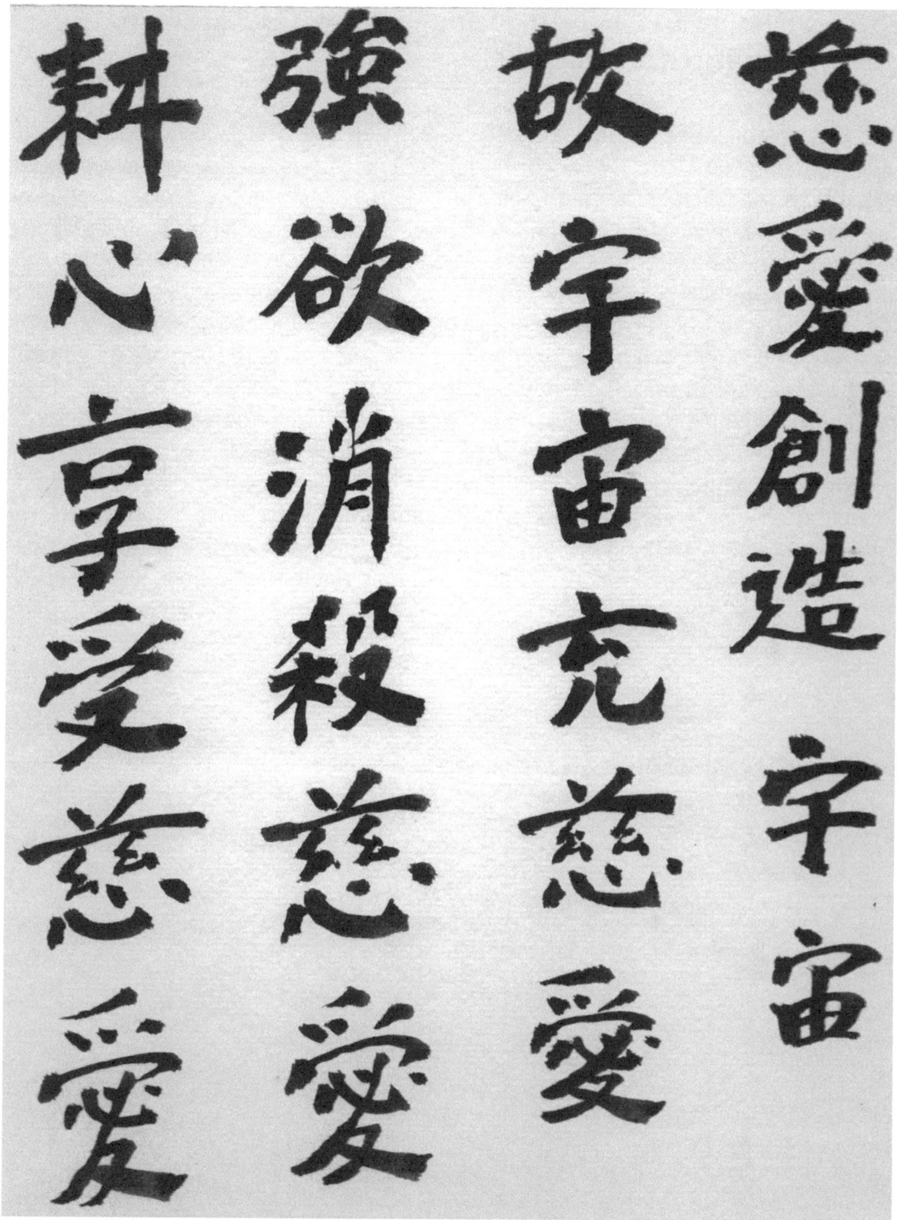

Fig. 15.1 A philosophical poem by the author. "Affection created the universe, thus universe is full of affection. Avarice eliminates and kills affection, a well cultivated mind enjoys affection"

15.2 From an Ill Concept "Time Is Money" to Attaining a Healthy Currency

The concept of "time is money" leads directly to individual competition and ill treatment of others, even wars. Economic human activities targeted at obtaining currency have resulted in environmental disorder. For example, the increase of CO_2 may result in catastrophic climate changes. Money is just a tool for maintaining the three stages of life, which is why there is no need for too much of it in order for a human being to live comfortably and receive affections from the universe (nature). Big concentrations of currency are dangerous, because they no longer represent only currency. A healthy currency is something that is utilized for the purpose of cultivating the third stage of life of individual children before they start attending school and of mature individuals as well as maintaining a healthy environment. If each of us discovered our personality for creations, we would all be respectful of each other, and we would consequently avoid "ill treatment of others" and further "wars."

Figure 15.1 indicates a philosophical poem in Chinese read by the author in 2009, meaning: "Affection created the universe, thus universe is full of affection. Avarice eliminates and kills affection, a well cultivated mind enjoys affection."

References

Ando Y (1985) Concert hall acoustics. Springer, Heidelberg

Ando Y (1998) Architectural acoustics, blending sound sources, sound fields, and listeners. AIP Press/Springer, New York

Ando Y (2007, 2014) Concert hall acoustics based on subjective preference theory. In: Rossing TD (ed) Springer handbook of acoustics, 1st, 2nd edn. Springer, New York

Ando Y (2009a) Auditory and visual sensations. Springer, New York

Ando Y (2009b) Theory of temporal and spatial environmental design. In: McGraw-Hill yearbook of science & technology 2009. McGraw-Hill, New York, pp 384–389

Appendices: Duration Experience and Cyclic Universe

A.1 A Model of Duration Experience in Listening to Noise and Music

A.1.1 Model of Duration Experience

To understand duration experience, a model of duration experience is first proposed here as shown in Fig. A1.1. Endogenous clock oscillators with characteristic periods can be synchronized with, for example, brain waves, pulsation, and breathing rate inside the body, in addition to daily, monthly, and annual cyclical changes in the external environment (Luce 1971). That is to say, morning light may reset a circadian oscillator, and a long-term temperature change may reset a seasonal oscillator. It is assumed that clock signals from oscillators propagate through a system (media) to a receptor in the brain.

Activities of the receptor in the brain depend on the person being awake or asleep. In the waking state, there are two levels of perception, "conscious" and "unconscious." In the conscious stage, a person always notices time passing, whereas in the unconscious state, they do not notice time passing. When a person is in deep sleep, for example, their receptor gets less endogenous clock signals and they may feel time passes faster than it does while they are awake.

We have attained knowledge of the effects of sound on subjective impressions of passing time. A question arises whether or not music and environmental noise influence unconscious duration experience.

Y. Ando, *Brain-Grounded Theory of Temporal and Spatial Design*,
DOI 10.1007/978-4-431-55891-0

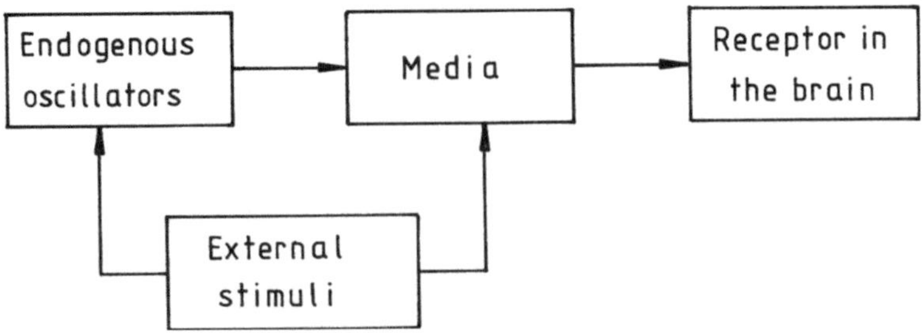

Fig. A1.1 A model of duration experience consisting of endogenous oscillators, media, and a receptor in the brain. According to the model of an auditory-brain system and specialization of cerebral hemispheres (Ando 1985, 1998), it is assumed that the receptor exists in the left hemisphere (Polzella et al. 1977). External stimuli may act as a suppressor to a number of clock pulses from the oscillators, so the receptor may get less clock pulses and the person may feel a short unconscious duration (Ando 1977)

A.1.2 Unconscious Duration Experience in Listening to Aircraft Noise

In the first hour of the experiment, picture slides of a world tour were shown in a classroom to make the subjects relax. During the following hour, cycles composed of 60 s of jet-plane noise followed by 60 s of silence were repeated 30 times. As listed in Table A1.1, the peak sound-pressure levels (SPL) in all the classrooms of the junior high school in the study were 75 ± 2 dBA (group A, 20 subjects; group B, 28 subjects) and 90 ± 2 dBA (group C, 20 subjects; group D, 25 subjects) at peak level. Group A and group C subjects lived in a noisy residential area, while group B and group D subjects lived in a quiet residential area. During the hour of noise-silence cycles, the subjects were permitted to look at picture books. For the last hour, the slide show was presented once more to make the students relax again. Afterward, the subjects were given a questionnaire asking for the ratio of subjective duration between the period of noise and the period of silence. That is, they were asked, "By how much was the noise period longer than the silence period?" The question was carefully explained so that all subjects understood the question and could answer it meaningfully, choosing a ratio from 0.1, 0.2, 0.3, . . . 5.0. Note that the objective ratio of time is unity, i.e., 60/60 s.

Results including the ratios (R_{50}) at 50 % for four groups are listed in Table A1.1. A notable result is shown in Fig. A1.2. About 79 % of group A and group B (75 ± 2 dBA) rated the subjective duration of the noise period shorter than that of the silent period ($p < 0.05$; the null hypothesis: 50 %). Mean values R_{50} of group A and group B, respectively, are 0.4 and 0.5. About 80 % of subjects from group C (90 ± 2 dBA) (from a noisy residential area) rated the subjective duration of the noise period somewhat shorter than the silent period ($p < 0.05$), i.e., $R_{50} = 0.7$.

According to the model shown in Fig. A1.1, a noise environment might suppress signals from endogenous clock oscillators such as from specific brain waves with characteristic periods, so that perceived duration is shorter. In particular, when the

Table A1.1 Number of subjects in four groups (pupils of junior high school), percentage of subjects who rated the duration of noise shorter than silence, and the value of R_{50}

Group	Residential area	SPL (dBA)	Number of subjects	Percent (%)	R_{50}	Null hypothesis (%)	p
A	Noisy area	75 ± 2	20	80	0.4	50 (R_{50} = 1.0)	<0.05
B	Quiet area	75 ± 2	28	78	0.5	50 (R_{50} = 1.0)	<0.05
C	Noisy area	90 ± 2	20	80	0.7	50 (R_{50} = 1.0)	<0.05
D	Quiet area	90 ± 2	25	52	0.9	50 (R_{50} = 1.0)	–

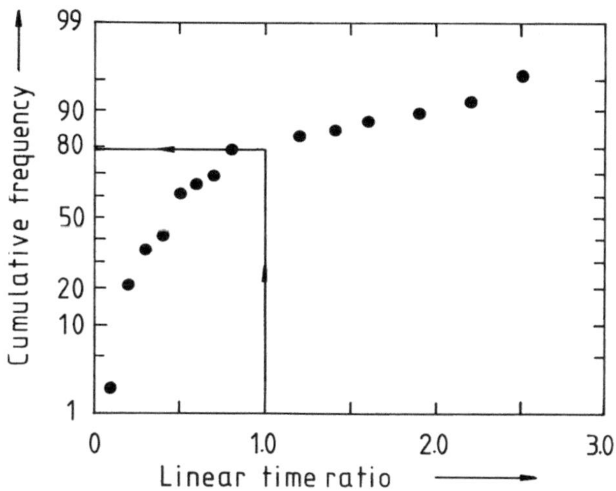

Fig. A1.2 Cumulative frequency in relation to the time ratio. Mean value R_{50} of group A is 0.4; time passes faster

sound-pressure level was about 75 dBA for subjects from a noisy residential area, the rate for about 80 % of the subjects became ($R_{50} = 0.4$). This result is considered to be a chronic effect of daily noise.

This result may relate to people living in a noisy area near an airport, who felt busy even though they were doing no particular job.

If we stretch the results obtained here, the duration of a life span of 100 years could be perceived as lasting only about 40 years in a noisy living area at about 75 dBA.

A.1.3 Duration Experience in Listening to Slow Music and Fast Music

The number of subjects that participated in this experiment was 102 (20–22 years old), all of whom were university students attending a class on environmental psychology and physiology (Ando and Ando 2010). To provide a condition for an unconscious duration experience while listening to music, the subjects were not acquainted with the real purpose of the study. The masker investigation was about subjective preference for a sound field with and without an echo, which consisted of a direct sound and a single reflection with a delay time of 128 ms (Ando 1985). The music sources were music A (a slow tempo "quiet" music piece, Royal Pavane by Gibbons; duration of piece, 9.26 s with a peak level of 70.5 ± 2 dBA; the effective duration of the autocorrelation function $\tau_e = 127$ ms) and music B (a fast tempo "active" music piece, Sinfonietta, Opus 48 by Arnold; duration, 6.26 s with a peak level of 84 ± 4 dBA; $\tau_e = 43$ ms). The values of τ_e represent a similar repetitive feature of music signals. The two music pieces were reproduced four times in the room, and the perceived loudness of one was almost the same as the other, though the peak levels were different. A questionnaire given to each subject posed two questions: the first was, "Do you prefer music A or music B?" The second question was, "How much longer was the duration of music A (Gibbons, quiet music) compared to music B (Arnold, active music)," choosing a "ratio of A to B" from 0.1, 0.2, 0.3, . . . 10.0 times. Note that the objective ratio of time was 1.48 = 9.26/6.26 s.

The percentage cumulative frequencies of two groups, namely, subjects who preferred music A (group A) and music B (group B), are plotted as a function of the time ratio in Fig. A1.3 (also see Table A1.2). About 70 % of group A rated the ratio of subjective duration as being greater than the physical ratio (1.48); that is, the subjective duration of music A was thought to be longer than that of music B ($p < 0.01$). The difference between duration experiences of group A and group B may be found in values expressed as $R_{50} = 2.0$ and $R_{50} = 1.5$ times ($p < 0.1$), respectively. This difference was found regardless of the value of R_x ($x = 10$ to 90), so curves of the cumulative frequency of both groups are parallel (Fig. A1.3). This result means that when the subjects preferred music B, the duration experience for music B was relatively prolonged.

Of particular interest is the difference in subjective preference for the two music pieces. More subjects in group A rated the ratio of subjective duration as being

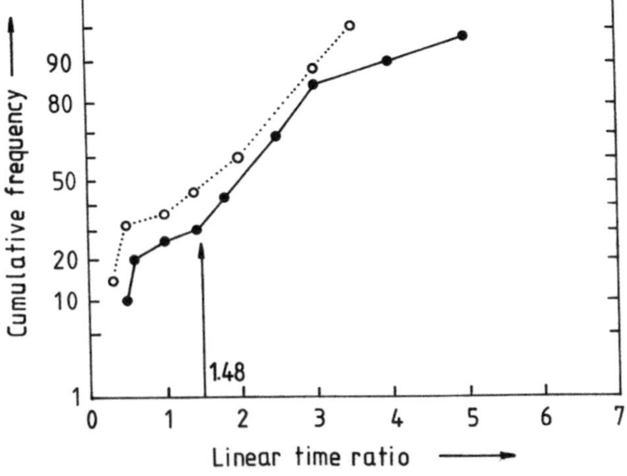

Fig. A1.3 Percentage cumulative frequencies of two groups of subjects classified by their preference as a function of the ratio of the duration experiences for the music A period and for the music B period
●: Music A, 80 subjects, who judged music A as preferable to music B
○: Music B, 22 subjects, who judged music B as preferable to music A

Table A1.2 Percentages of subjects (university students) who rated the duration of music A longer than the physical ratio (1.48 times) of music B

Group classified by preference	Number (valid) of subjects	Percent (%)	R_{50}	Null hypothesis (%)	p
Music A	80	70	2.0	50 ($R_{50} = 1.48$)	<0.01
Music B	22	50	1.5	50 ($R_{50} = 1.48$)	–

greater than the physical ratio (1.48); that is, the subjective duration of music A was significantly longer than that of music B. It is considered that active music B may suppress the endogenous clock signals more than the quiet music (Fig. A1.1). Additionally, when subjects preferred music B, their duration experience of music B increased relatively. This implies that subjects liked listening to a preferred piece of music over a longer time period, because it induced more alpha brain waves (which are the longest brain waves in the awake state) to enjoy it for as long as possible. This, in turn, might prolong the period of the endogenous clock pulse.

It is noteworthy that the similar repetitive features of the alpha wave are prolonged under the preferred condition, which has been determined by the effective duration of the autocorrelation function of the alpha wave (Ando 2003, 2009).

With regard to joyful artistic and scientific creations, a receptor is a typical unconscious level that might originate from "out of time" or beyond any specified time (Ando 2009). Therefore, such creation based on individual personality might relate to an individual's "third stage of life" that lives on even after the end of the first (body) and second (mind) lives (Sect. 1.2; Ando 2004).

A.2 Duration Experience for Each Three-Year School Term

A tendency was found that the subjective time duration experience of every 3-year term during senior high school was much shorter than the perceived temporal experience during elementary school in reference to the 3 years of junior high school. This result agreed with the well-known perception of time passing faster as one ages.

The Japanese education system that is based on 3-year terms was investigated for subjective impressions covering 3-year "blocks" of the 15-year-long school life (from about 3 to 18 years of age). In order to minimize the different effects of present stage life on each individual, university students on ocean navigation for 30 days between Kobe, Japan, and Perth, Australia, in a summer session in 1994 were selected as subjects (Ando et al. 1999). The total number of subjects who participated in this investigation that was performed in a large lecture room of a chartered boat was 285 (81 % female and 19 % male). The majority of the students were juniors and sophomores.

Questionnaires were distributed during the voyage, and the subjects were asked to carefully respond to questions about their subjective impression of the following time durations: in reference to the subjective time duration of the junior high school period of 3 years (JHSP), (1) duration of 3 years before elementary school, (2) duration of 3 years of lower class elementary school, (3) duration of 3 years of upper class elementary school, and (4) duration of 3 years of senior high school. The method used was similar to the magnitude estimation, so that subjective durations were answered choosing one of the real numbers from 0.1 through 3.0 with a step of 0.1, and in addition, the subjects were allowed to answer with other numbers as well.

For subjective duration with (1) gender of subject, (2) experience of transfer between elementary schools due to change of the family's residence, and (3) period of every 3 years, the analysis of variance was performed. Significant differences were found in the 3-year periods as well as in the experience of transfer between elementary schools, but not for the gender of the subject. Since the subjective duration estimate is well described by time on the logarithmic scales, meaningful mean values may be found at the 50th percent of cumulative frequencies. Figure A2.1 indicates the 50-percentile values as a function of the period of the 3-year range, for both subjects who moved and changed elementary schools and for subjects who did not.

For subjects who did not transfer between elementary schools, results from 3 to 5 years of age through 9–11 years of age reveal that subjective durations are longer than unity, 1.2, and results from 15 to 17 years of age indicate that subjective impression for 15–17-year-olds are significantly shorter than unity, 0.8, in reference to JHSP ($p < 0.01$). Thus, a tendency is observed that the duration experience decreases with increasing age. For subjects who changed elementary schools, the subjective durations for 3–5 and 6–8 years of age were significantly longer, about 1.3 or more, than for subjects who did not transfer between schools ($p < 0.05$) due to additional strong stress of having to adapt to a new environment.

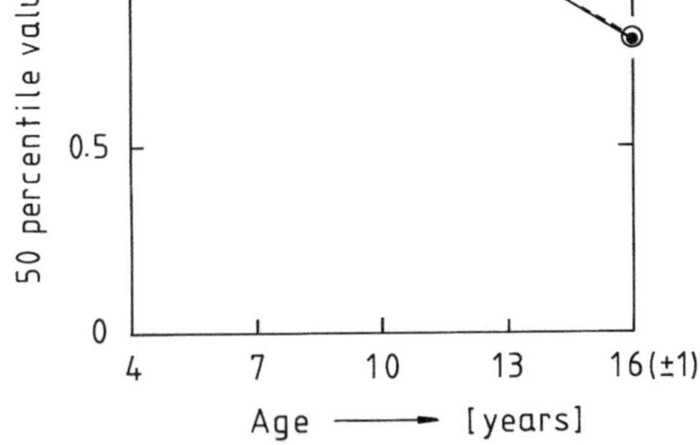

Fig. A2.1 The 50-percentile values of duration in reference to duration of 3 years of junior high as a function of the period of the 3-year range, for both subjects who moved and changed elementary schools (*empty circle*) and for subjects who did not (*full circle*)

A.3 Longest Period: Cyclic Universe

In 1953, Asano Ando (1906–1996; mother of the author) was witnessing a forest fire across a river in Mimasaka, Okayama Prefecture, when she expressed her hypothesis that the whole universe has ended and started again repeatedly. She continued by saying that this world would not be appropriate for life if such fires occurred time and again in the future.

There could be a certain relationship between the environment from before a big bang and after one, i.e., indications of a former universe are assumed by the author. Typical material is organic matter in the form of amino acids, which were first discovered in a meteorite by Kvenvodden et al. 1970. Could we consider this as a kind of continuation and reproduction of the environment in a cyclic universe? Who is to say which parts or dimensions of the universe are newly produced and which ones might survive a cataclysm such as a big crunch or a big bang?

Steinhardt (2009) has explained the theory of the cyclic universe, which was inspired by developments in string theory. A model of cosmic evolution of the universe is believed to undergo endless cycles of expansion and contraction, each beginning with a big bang and ending in a big crunch. Thus, the bangs have occurred periodically in the past and will continue periodically in the future, recurring perhaps once every 10^{12} years. Such a periodical universe consists mainly of the four dimensions of time and space, and other dimensions could be ignored.

References

Ando Y (1977) Effects of noise on duration experience. J Sound Vib 55:600–603
Ando Y (1985) Concert hall acoustics. Springer, Heidelberg
Ando Y (1998) Architectural acoustics, blending sound sources, sound fields, and listeners. AIP Press/Springer, New York
Ando Y (2003) Investigations on cerebral hemisphere activities related to subjective preference of the sound field, published for 1983–2003. J Temporal Des Architect Environment 3:2–27, http://www.jtdweb.org/
Ando Y (2004) On the temporal design of environments. J Temporal Des Architect Environ 4:2–14. http://www.jtdweb.org/
Ando Y (2009) Auditory and visual sensations. Springer-Verlag, New York
Ando Y, Ando T (2010) Model of "unconscious" duration experience while listening to music and noise. J Temporal Des Architect Environ 10:1–6, http://www.jtdweb.org/
Ando Y, Tsuruta H, Motokawa A, Matsushita T, Saifuddin K (1999) Subjective duration of every three-year period for 3 to 18 years of age, estimated by students. J Hum Ergol 28:33–37
Kvenvodden KA, Lawless J, Pering K, Peterson E, Flores J, Ponnamperuma C, Kaplan IR, Moore C (1970) Evidence for extraterrestrial amino-acids and hydrocarbons in the Murchison meteorite. Nature 228:923–926
Luce GG (1971) Biological rhythms in human and animal physiology. Dover Publications, New York
Polzella DJ, DaPolito F, Hinsman MC (1977) Cerebral asymmetry in time perception. Percept Psychophys 21:187–192
Steinhardt PJ (2009) Cyclic universe theory. McGraw-Hill Yearbook of Science & Technology, New York, pp 69–71

Index

Y. Ando, *Brain-Grounded Theory of Temporal and Spatial Design*,
DOI 10.1007/978-4-431-55891-0

MIX
Papier aus verantwortungsvollen Quellen
Paper from responsible sources
FSC® C105338

If you have any concerns about our products,
you can contact us on
ProductSafety@springernature.com

In case Publisher is established outside the EU,
the EU authorized representative is:
Springer Nature Customer Service Center GmbH
Europaplatz 3, 69115 Heidelberg, Germany

Printed by Libri Plureos GmbH
in Hamburg, Germany